Math Mammoth
Grade 6 Review Workbook

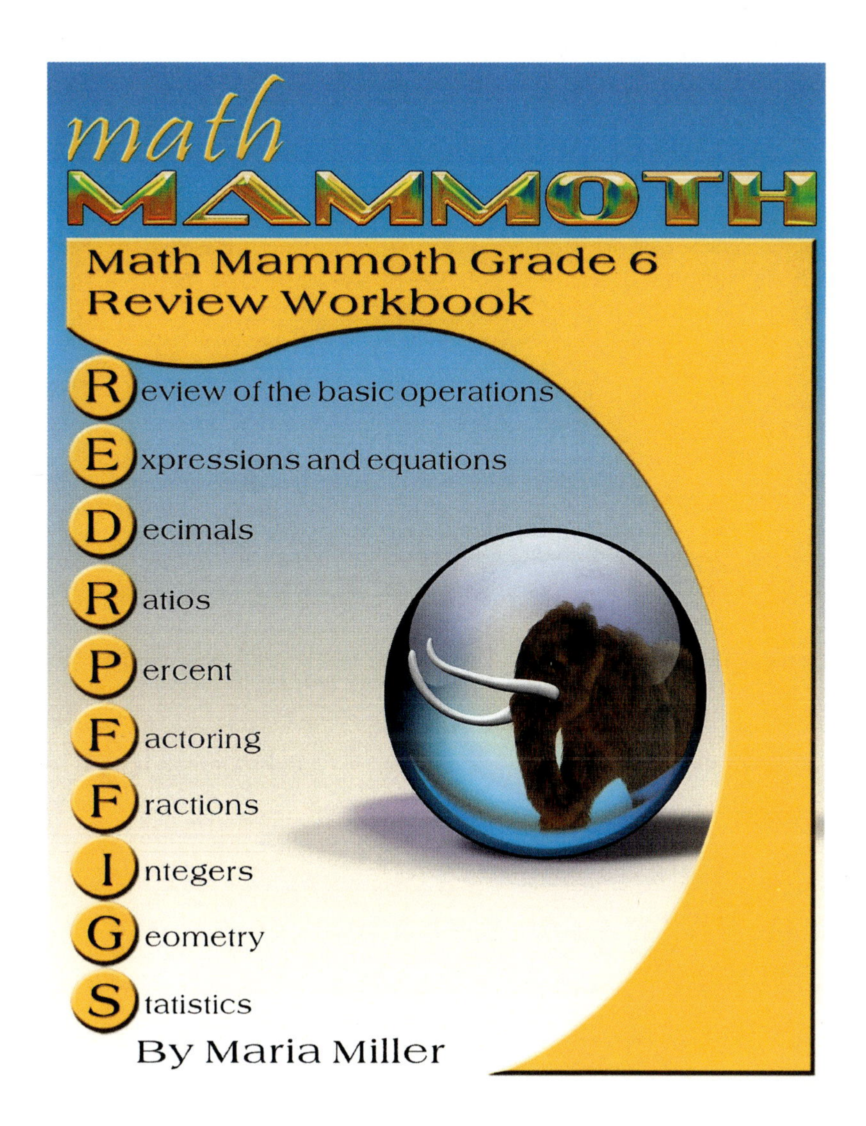

By Maria Miller

ISBN 978-1516844081

EDITION 8/2015

Math Mammoth Grade 6 Review Workbook
Contents

Introduction

Math Mammoth Grade 6 Review Workbook is intended to give students a thorough review of sixth grade math, following the main areas of Common Core Standards for grade 6 mathematics. The book has both topical as well as mixed (spiral) review worksheets, and includes both topical tests and a comprehensive end-of-the-year test. The tests can also be used as review worksheets, instead of tests.

You can use this workbook for various purposes: for summer math practice, to keep the child from forgetting math skills during other break times, to prepare students who are going into seventh grade, or to give sixth grade students extra practice during the school year.

The topics reviewed in this workbook are:

- review of the four basic operations
- expressions and equations
- decimals
- ratios
- percent
- factoring
- fractions
- integers
- geometry
- statistics

In addition to the topical reviews and tests, the workbook also contains many cumulative (spiral) review pages.

The content for these is taken from the *Math Mammoth Grade 6 Complete Curriculum*, so this workbook works especially well to prepare students for grade 7 in Math Mammoth. However, the content follows a typical study for grade 6, so this workbook can be used no matter which math curriculum you follow.

Please note this book does not contain lessons or instruction for the topics. It is not intended for initial teaching. It also will not work if the student needs to completely re-study these topics (the student has not learned the topics at all). For that purpose, please consider the *Math Mammoth Grade 6 Complete Curriculum*, which has all the necessary instruction and lessons.

I wish you success with teaching math!

Maria Miller, the author

The Basic Operations Review

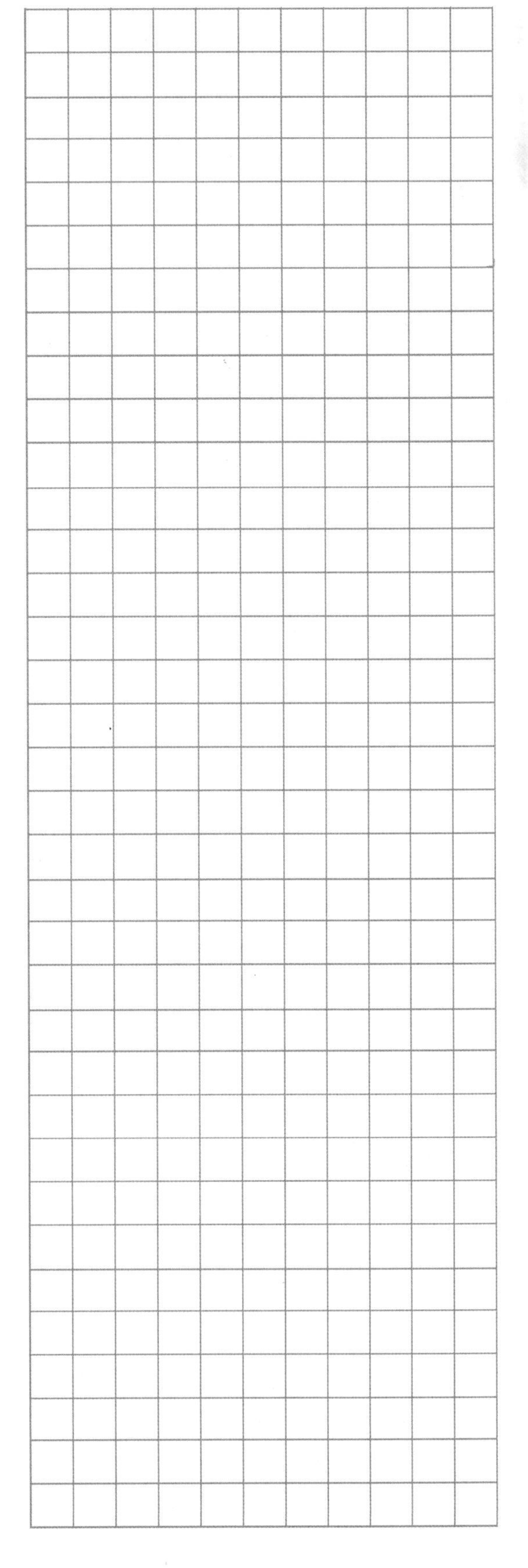

1. Divide and indicate the remainder, if any. Use long division.

   a. 6,764 ÷ 81

   b. 309,855 ÷ 46

2. How many times can you subtract 9 from 23,391 before you "hit" zero?

3. If you spend exactly $2.25 every day to make a phone call, how much will those phone calls cost you in a year?

4. If 5,000 people need to be moved from place A to place B by buses, and one bus seats 46 people, how many buses are needed?

5. An airplane travels at a constant speed of 880 km per hour. *Estimate* about how long it will take for it to fly 5,800 km.

6. Three boxes of tea bags cost $15.90.
How much do two boxes cost?

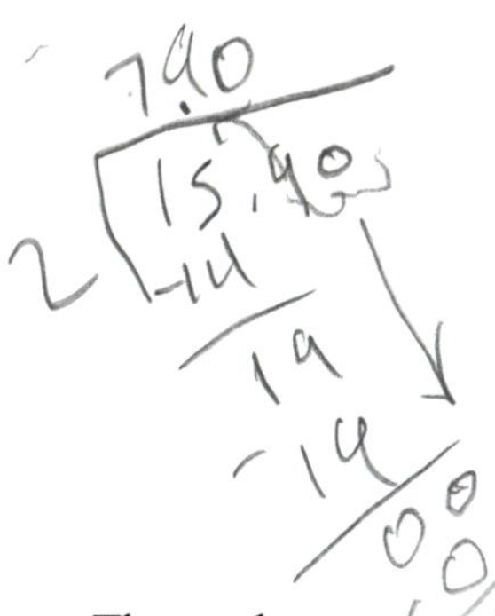

7. Write the expressions using an exponent. Then solve.

a. $5 \times 5 \times 5 \times 5$

b. $1 \times 1 \times 1 \times 1 \times 1 \times 1$

c. 30 squared

d. $100 \times 100 \times 100$

e. two to the sixth power

f. three cubed

8. **a.** The perimeter of a square is 80 cm. What is its area?

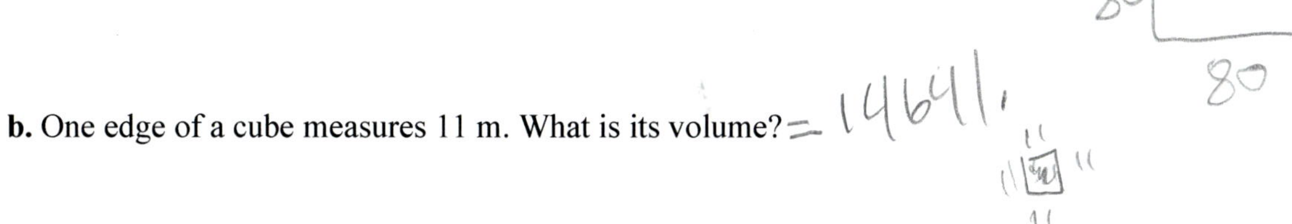

b. One edge of a cube measures 11 m. What is its volume?

9. Fill in.

a. 25^3 gives us the ____________ of a ______________ with edges ______ units long.

b. 3×9^2 gives us the ____________ of ______ ______________ with sides ______ units long.

10. Write in normal form (as a number).

a. $2 \times 10^5 + 3 \times 10^2 + 9 \times 10^0$

b. $2 \times 10^7 + 8 \times 10^6 + 3 \times 10^4 + 1 \times 10^3$

11. Write in order from the smallest to the largest.

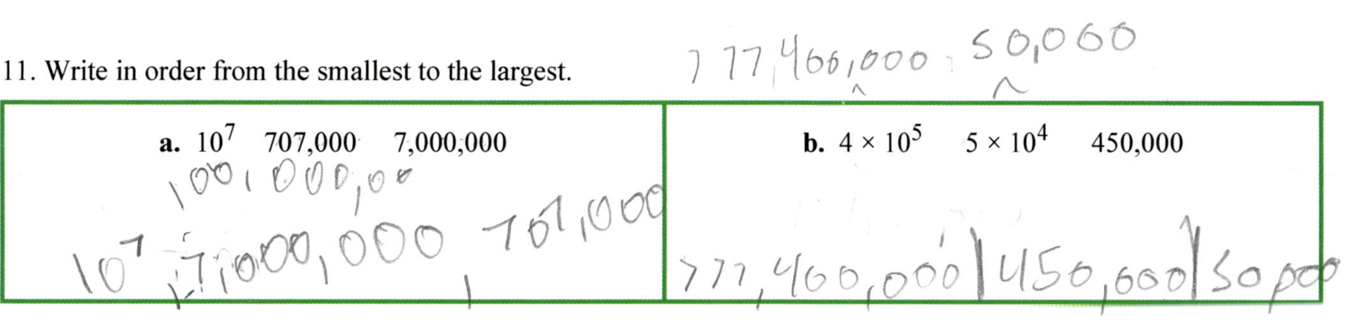

a. 10^7 707,000 7,000,000	**b.** 4×10^5 5×10^4 450,000

12. Round to the place of the underlined digit. Be careful with the nines!

a. 14<u>9</u>,601 ≈ ______________

b. 2,9<u>9</u>9,307 ≈ ______________

c. 59<u>7</u>,104,865 ≈ ______________

d. 559,9<u>9</u>8,000 ≈ ______________

The Basic Operations Test

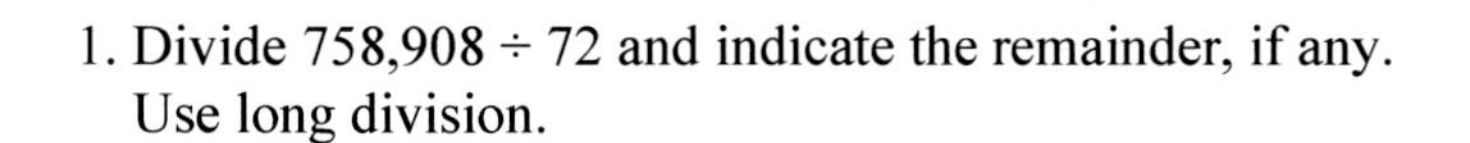

1. Divide 758,908 ÷ 72 and indicate the remainder, if any. Use long division.

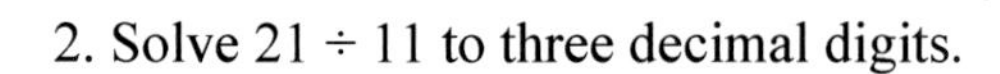

2. Solve 21 ÷ 11 to three decimal digits.

3. A box of 75 flashlights costs $937.50. Find the cost of one flashlight.

4. You can scan one page of a book in 43 seconds. Working at the same speed, how long will it take you to scan all 234 pages of the book?
   Give your answer in minutes and seconds.

5. Solve.

   a. 3^3

   b. 1^{10}

   c. 50^2

   d. 10^5

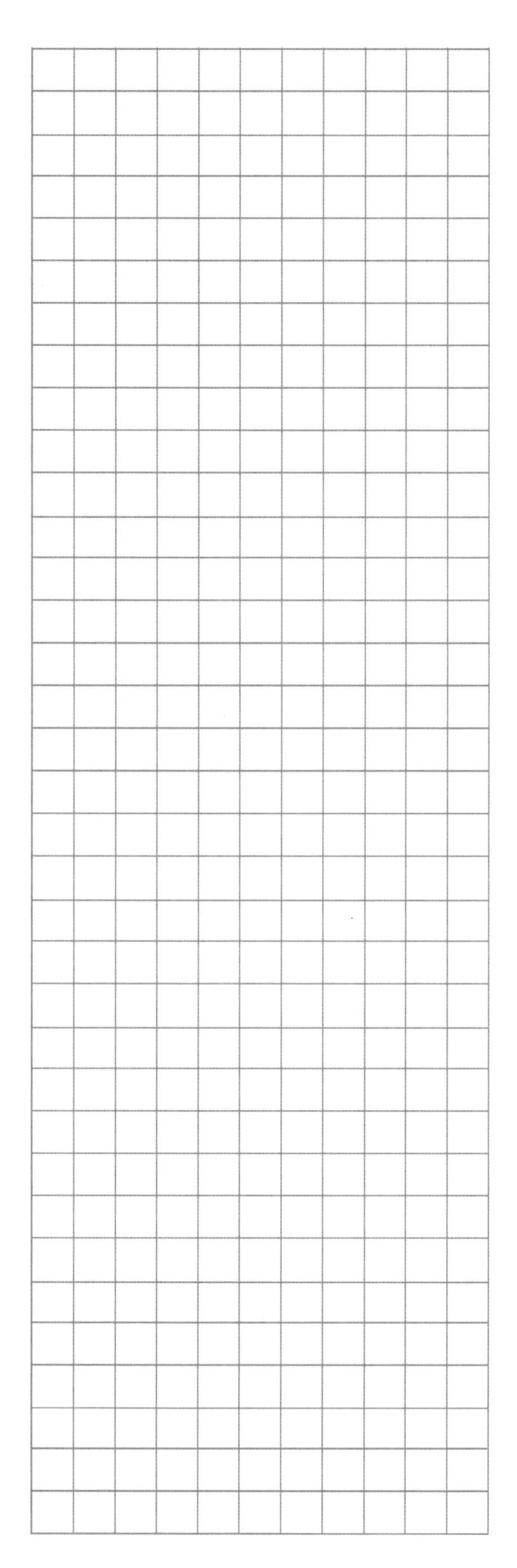

6. The perimeter of a square is 56 cm. What is its area?

7. Write in normal form (as a number).

a. $5 \times 10^8 + 4 \times 10^6 + 3 \times 10^5$

b. $1 \times 10^9 + 6 \times 10^8 + 2 \times 10^4 + 1 \times 10^2$

8. Write in expanded form, using exponents (as in the original in #7).

a. 560,000

b. 9,108,000

9. Round the numbers.

a. 2,998,601 to the nearest ten thousand

b. 483,381,902 to the nearest ten million

c. 19,993,740 to the nearest million

Expressions and Equations Review

1. Write an expression.

   a. the difference of 6 and x, squared

   b. the quotient of 5 and the sum of x and 6

   c. 3 times the quantity 5 minus p

2. Find the value of the expressions.

a. $(1 + 6)^2 + (10 - 2)^2$	**b.** $5^2 \cdot 2^3$
c. $\dfrac{21 + 6}{2 \cdot 1 + 1}$	**d.** $\dfrac{16}{2} \cdot (120 - 50)$

3. Find the value of the expressions.

a. $2x + 18$ when $x = 5$	**b.** $\dfrac{35}{z} \cdot 13$ when $z = 5$

4. Write an expression for each situation.

   a. Three friends purchased a scuba diving outfit together for p dollars. They shared the cost equally. How much did each person pay?

   b. You bought modeling clay for $3 and six packages of crayons for c dollars each. What was the total cost?

5. Label each thing below as an equation, inequality, or expression.

   $2x + 17$ $8 = 8$ $y < 5$ $4x - 3 = 8$ $\dfrac{4}{5}x - 16$ $4x + y^2 \geq 9$ $M = \dfrac{44 - x}{5}$

6. Simplify the expressions.

a. $t + t + t + 3$	**b.** $8d - 3d$
c. $x \cdot x \cdot x$	**d.** $12x - 6 - 6x$
e. $z \cdot z \cdot 8 \cdot z \cdot 2$	**f.** $3x^2 + 5 + 11x^2$

7. Write an expression for *both* the area and perimeter of each rectangle. Give them in simplified form.

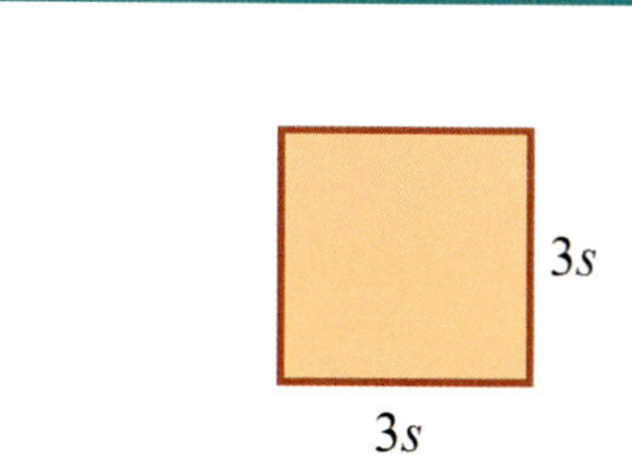

a. A =

P =

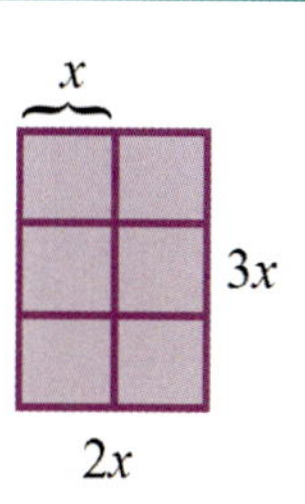

b. A =

P =

8. Multiply using the distributive property.

a. $3(2x + 7) =$	**b.** $8(9b + 5) =$

9. Factor these sums, thinking of the distributive property "backwards."

a. $5x + 10 =$ ____ $(x +$ ____ $)$	**b.** $6y + 10 =$ ____ (____ + ____)
c. $24b + 4 =$ ____ (____ + ____)	**d.** $25w + 40 =$ ____ (____ + ____)

10. Solve the equations.

a. $7x = 784$	**b.** $3 + z = 119$	**c.** $\frac{x}{6} = 12$
d. $5y + 8y = 784$	**e.** $32 + x = 9 \cdot 40$	**f.** $\frac{r}{6 + 4} = 7$

11. Write an equation for each situation even if you could easily solve the problem without an equation. Then solve the equation.

a. The value of a certain number of quarters is 1675 cents. How many quarters are there?

b. The perimeter of a rectangle is 128 meters. One side is 21 meters. How long is the other side?

12. The formula $F = \frac{9C}{5} + 32$ is used to convert temperatures given in Celsius degrees into Fahrenheit degrees.

C denotes the temperature in Celsius degrees, and F denotes the temperature in Fahrenheit. If the temperature in Celsius is 25°C (nice summer weather), find the corresponding temperature in Fahrenheit.

13. Write an inequality that corresponds to the number line plot.

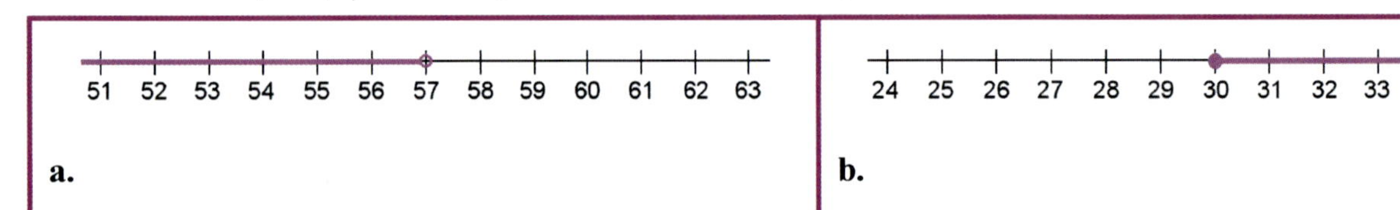

14. **a.** Solve the inequality $y + 2 > 24$ in the set {55, 44, 22, 23, 30}.

b. What solutions does the inequality $x + 7 \leq 14$ have in the set of even whole numbers?

15. Calculate the values of y according to the equation $y = x + 3$.

x	1	2	3	4	5	6
y						

Now, plot the points.

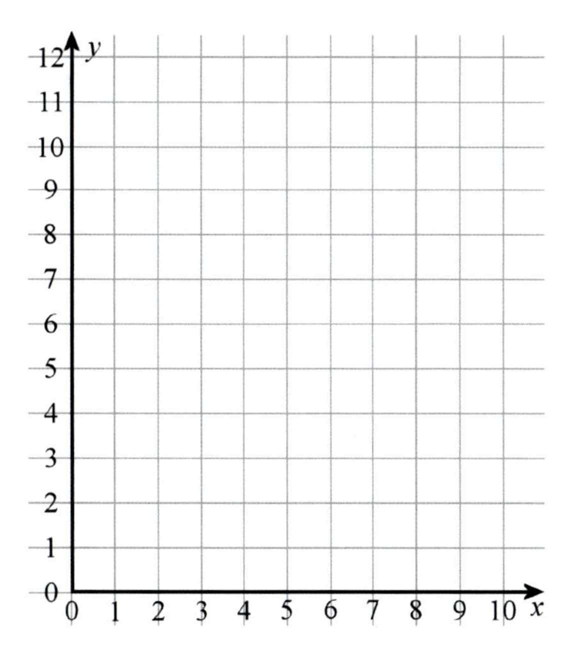

16. A train is traveling at a constant speed of 70 miles per hour. Consider the variables of time (t), measured in hours, and the distance traveled (d), measured in miles.

a. Fill in the table.

t (hours)	0	1	2	3	4	5	6
d (miles)							

b. Plot the points on the coordinate grid.

c. Write an equation that relates t and d.

d. Which of the two variables is the independent variable?

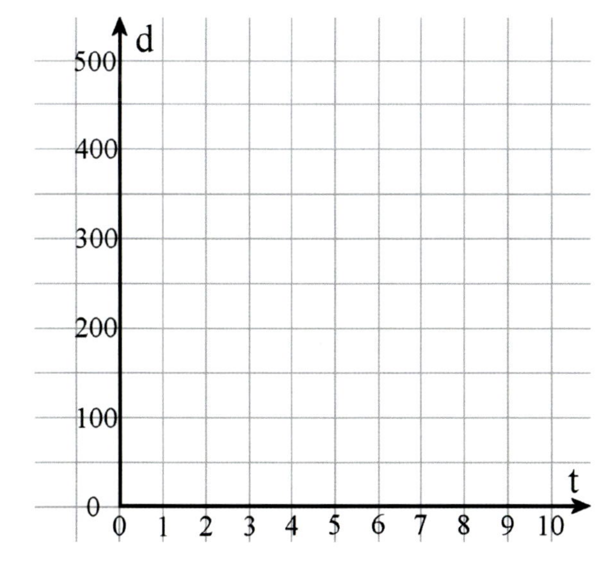

Expressions and Equations Test

1. Write an expression.

a. The quotient of x squared and 7.

b. The quantity 5 minus y, cubed.

c. 3 times the quantity $2s$ minus 5.

2. Find the value of these expressions.

a. $(100 - 80) \cdot 2 - 20$	**b.** $480 \div 2^3$	**c.** $32 + 0^5 \cdot 12 \div 4$

3. Find the value of the expressions.

a. $2x + 10$ when $x = 5$	**b.** $x^2 + 10$ when $x = 5$
c. $\dfrac{40 - x}{5}$ when $x = 5$	**d.** $40 - \dfrac{5}{x}$ when $x = 5$

4. Write an expression.

You purchase a book for p dollars and three pencil cases for t dollars each. What is your total cost?

5. Simplify the expressions.

a. $a \cdot a \cdot a \cdot a$	**b.** $a + a + a + a$
c. $x \cdot x \cdot 5 \cdot 2$	**d.** $8d - 2d + 7$

6. Multiply using the distributive property.

a. $5(x + 6) =$	**b.** $2(9 + 5y) =$

7. Solve the equations.

a. $6x = 144$	**b.** $y + 78 = 134$	**c.** $\dfrac{x}{16} = 3$

8. Write an equation <u>even if</u> you could easily solve the problem without an equation! Then solve the equation.

The perimeter of a square is 164 units. How long is its side?

9. Plot these inequalities on the number line.

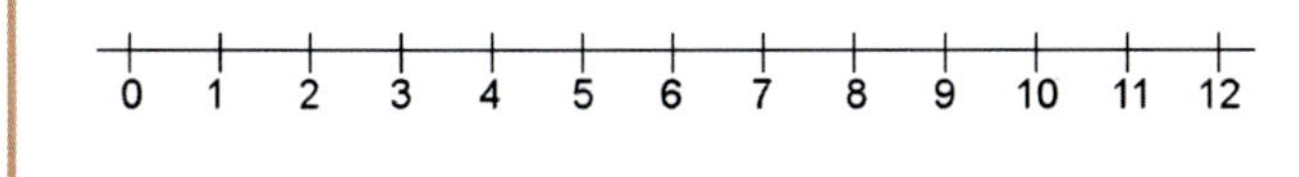

a. $x \geq 5$

0 1 2 3 4 5 6 7 8 9 10 11 12

b. $x < 8$

10. Solve the inequality $x + 3 < 20$ in the set {15, 16, 17, 18, 19, 20}.

11. Calculate the values of y according to the equation $y = 9 - x$.

x	0	1	2	3	4	5	6	7	8	9
y										

Now, plot the points.

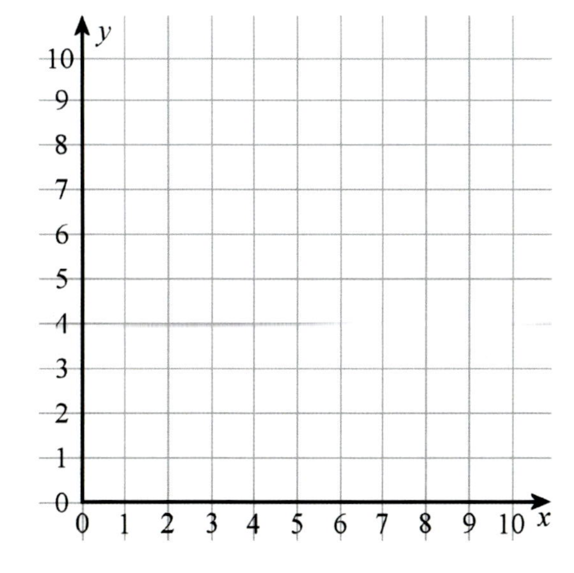

Mixed Review 1

1. Solve.

a. 10^5	**b.** 3^4	**c.** $10^5 \cdot 4$

2. Which power of ten is equal to ten million?

3. Express the area (A) or volume (V) as a multiplication, and solve.

a. A square with a side of 3 kilometers: A = ______________________	**b.** A cube with edges that are 2 inches long: V = ______________________

4. **a.** The area of a square is 81 cm^2. What is its perimeter?

   b. The perimeter of a square is 48 m. What is its area?

5. Write as numbers.

   a. 5 trillion, 51 billion, 27 thousand

   b. 21 trillion, 650 billion, 99 million, 56

6. Write in normal form (as a number):

   $6 \cdot 10^6 + 2 \cdot 10^3 + 1 \cdot 10^0$

7. Write in expanded form using exponents.

   a. 54,000

   b. 2,090,030

8. Write an expression.

   a. The quotient of 5*s* and 8.

   b. 7 times the quantity *x* plus 8.

   c. *y* less than 8.

   d. The quantity *x* minus 8, squared.

9. Estimate the result using mental math and rounded numbers. Find the exact value using a calculator. Also, find the error of estimation.

a. $591 \cdot 57{,}200$ Estimation: Exact: Error of estimation:	**b.** $435{,}212 + 9{,}319{,}290$ Estimation: Exact: Error of estimation:

10. Solve. Notice carefully which operation(s) are done first.

a. $4 \cdot 50 + \dfrac{310}{2} =$ __________	**b.** $\dfrac{4{,}800}{60} - (70 - 20) =$ __________

11. A bicycle has been discounted by 2/10 of its price, and now it costs $120. Find the price before discount.

12. Divide. Use the space on the left to build a multiplication table for the divisor. Lastly, check.

	$79\overline{)\,5\;6\;2\;7\;9\;0}$	$\cdot\ \ 7\ 9$ ______

Mixed Review 2

1. **a.** A boat is traveling at the constant speed of 24 kilometers per hour. Fill in the table.

Distance		12 km		24 km		216 km
Time	10 min		50 min		5 1/2 hours	

b. How long will it take for the boat to travel 360 kilometers?

2. **a.** Estimate the answer to $234 \times 1{,}091$.

b. Now multiply $234 \times 1{,}091$ (in the space on the right, or in your notebook).

c. Next, estimate the answer to 2.34×1.091.

d. Lastly, based on your answers to (b) and (c), what is 2.34×1.091?

3. **a.** Write a subtraction equation where the minuend is 56, the difference is 17, and the subtrahend is the unknown y. Then solve for y.

b. Write a division equation where the quotient is 60, the divisor is 15, and the dividend is unknown. Solve it.

4. Divide mentally in parts.

a. $\frac{636}{6}$ **b.** $\frac{824}{4}$ **c.** $\frac{5{,}607}{7}$ **d.** $\frac{1{,}224}{12}$

5. Find the value of these expressions.

a. $100 - 100 \div 4 \cdot 2$	**b.** $3^3 \div (4 + 5)$
c. $(2 + 6)^2 - (25 - 5)$	**d.** $\frac{12^2 + 9}{5 \cdot 3}$

6. Evaluate the expressions when the value of the variable is given.

a. $3x - 12$ when $x = 5$	**b.** $\frac{y}{3} + 4$ when $y = 24$

7. Let w and l be the width and length of a rectangle. Which expression tells us the perimeter of the rectangle?

a. lw **b.** $2l + 2w$ **c.** $\frac{l}{w}$ **d.** $l + w + l$ **e.** $l + w$

8. Write an expression for the area (A) or volume (V) using an *exponent*, and solve.

a. A square with sides 11 cm in length: A = ______________________	**b.** A cube with edges that are all 4 ft long: V = ______________________

9. The perimeter of a square is 64 cm. What is its area?

10. One pair of shoes costs $48.60, and another pair costs 2/3 of that price.
Alyssa bought both pairs. Find her change from $100.

11. Divide. If the division is not exact, give your answer to three decimals.

a. $17 \overline{)2\,6\,7\,0\,8\,7}$	**b.** $15 \overline{)8}$	**c.** $3 \overline{)0.1\,3}$

Decimals Review

1. Write as decimals.

a. three ten-thousandths

b. 39234 hundred-thousandths

c. 4 millionths

d. 2 and 5 thousandths

2. Write as fractions.

a. 0.00039

b. 0.0391

c. 4.0032

3. Write as decimals.

a. $\frac{3}{4}$	**b.** $1\frac{2}{5}$	**c.** $\frac{17}{20}$	**d.** $\frac{11}{25}$

4. Fill in the table, noting that 1 micrometer is 1 millionth of a meter ($\frac{1}{1{,}000{,}000}$ of a meter)

Organism	Size (fraction)	Size (micrometers)	Size (decimal)
amoeba proteus	$\frac{600}{1{,}000{,}000}$ meters	_______ micrometers	0.0006 m
protozoa	from $\frac{10}{1{,}000{,}000}$ to $\frac{50}{1{,}000{,}000}$ m	from __10__ to __50__ micrometers	from _____ to _____ m
bacteria	from $\frac{1}{1{,}000{,}000}$ to $\frac{5}{1{,}000{,}000}$ m	from ______ to ______ micrometers	from _____ to _____ m

5. Write in order from the smallest to the largest.

a. 0.0256 0.000526 0.0062	**b.** 0.000087 0.000007 0.00008

6. Round to...

	0.37182	0.04828384	0.39627	0.099568
the nearest hundredth				
the nearest ten-thousandth				

7. Calculate mentally.

a. $0.02 + \frac{4}{1000}$	**b.** $0.7 + \frac{5}{100}$	**c.** $3.021 + \frac{22}{1000}$

8. Calculate. Remember to line up the decimal points.

a. $2.1 - 1.09342$

b. $17 + 93.1 + 0.0483$

9. Find the value of the expression $y + 0.04$ when

a. $y = 0.1$	**b.** $y = 0.01$	**c.** $y = 0.0001$

10. Divide mentally. For each division, write a corresponding multiplication.

a. $0.48 \div 6 =$	**b.** $1.5 \div 0.3 =$	**c.** $0.056 \div 0.008 =$

11. Multiply mentally.

a. $3 \times 0.006 =$	**b.** $0.2 \times 0.6 =$	**c.** $0.9 \times 0.0007 =$

12. 327×4 is 1,308. Based on that, figure out the answer to 32.7×0.004.

13. **a.** Estimate the answer to 8.9×0.061.

b. Calculate the exact answer.

14. Solve the equations by thinking logically.

a. $3 \times$ ________ $= 0.09$	**b.** $0.2 \times$ ________ $= 0.024$	**c.** $0.03 \times$ ________ $= 0.0015$

15. Solve the equations.

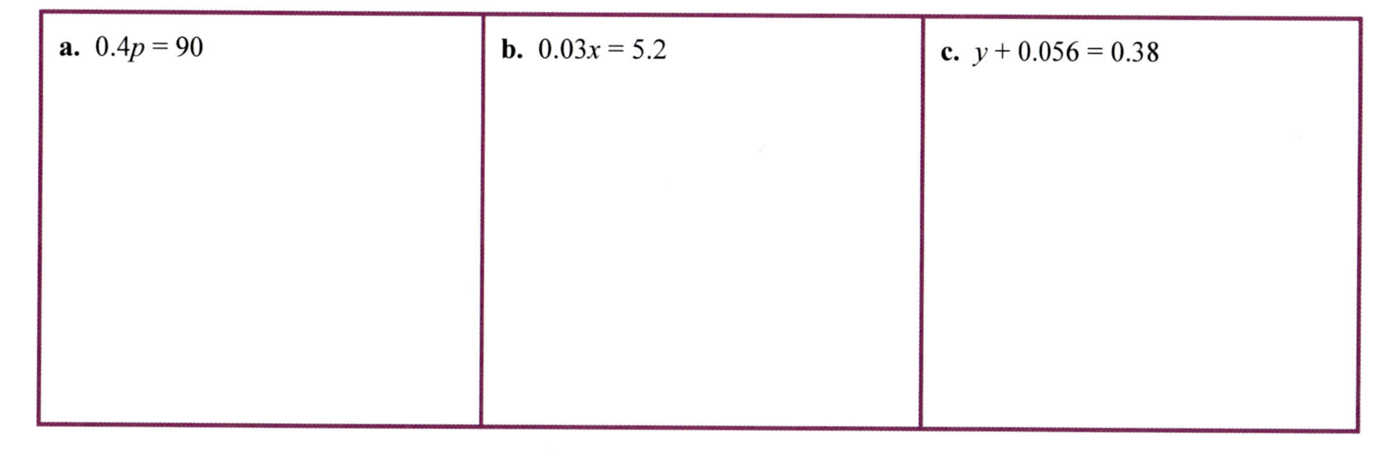

a. $0.4p = 90$	**b.** $0.03x = 5.2$	**c.** $y + 0.056 = 0.38$

16. Jim cut seven 0.56-meter pieces out of a 4-meter board.
How much is left?

17. Multiply or divide the decimals by the powers of ten.

a. $10^6 \times 21.7 =$	**b.** $100 \times 0.00456 =$
c. $2.3912 \div 1{,}000 =$	**d.** $324 \div 10^5 =$
e. $10^5 \times 0.003938 =$	**f.** $0.7 \div 10^4 =$

18. Find the value of the expression $\dfrac{a}{b} + 1$
when $a = 2.068$ and $b = 0.8$.

19. Divide, giving your answer as a decimal. If necessary, round the answers to three decimal digits.

a. $28.2 \div 2$	**b.** $0.11 \div 15$
c. $\dfrac{4}{9}$	**d.** $\dfrac{5}{11}$

20. Fill in the entries missing from this table.

Prefix	Meaning	Units - length	Units - mass	Units - volume
			centigram (cg)	
deci-				deciliter (dl)
	ten = 10		decagram (dag)	
				hectoliter (hl)

21. Change into the basic unit (meter, liter, or gram). Think of the meaning of the prefix.

a. 34 dl **b.** 89 cg **c.** 16 kl

22. Convert the measurements into the given units.

a. 2.7 L = _______ dl = ________ cl = ________ ml

b. 5,600 m = ________km = _________ dm = _________ cm

c. 676 g = _______ dg = ________ cg = ________ mg

23. You have eleven empty pop bottles. Six are 350 ml, two are 2 liters, and three are 9 dl. What is the total amount of water that you can put into them?

24. Convert into the given units. Round your answers to 2 decimals if needed.

a. 56 oz = ____________ lb	**c.** 2.7 gal = ___________ qt	**e.** 0.48 mi = ___________ ft
b. 134 in = ____________ ft	**d.** 0.391 lb = ___________ oz	**f.** 2.45 ft = _____ft _______ in

25. For a parade, each of 230 children needs a ribbon that is at least 2 feet long. If you buy a 500-ft roll of ribbon, how long (in feet and inches) will the ribbons be if you divide the roll equally?

26. One yard is 0.9144 meters. Which is a better deal:
40 yards of rope for $15.99
or 100 meters of rope for $40?

27. A scientist measured the length of some tadpoles caught from a pond. The recorded lengths are below, in centimeters. Find the average length of the tadpoles.
3.2 3.1 3.4 3.1 3.5 2.9 2.7 2.7 3.0 3.0 3.1
3.4 3.2 2.8 2.8 2.9 3.6 3.4 2.9 3.4 3.1

Decimals Test

1. Write as decimals.

a. five thousandths	**b.** 382 hundred-thousandths
c. 1 and 3,658 millionths	**d.** 94 ten-thousandths
e. $\frac{13}{20}$	**f.** $8\,\frac{2}{25}$

2. Write as fractions.

a. 2.0045	**b.** 0.000912	**c.** 7.49038

3. Calculate without a calculator.

a. $0.2 + \frac{5}{1000}$	**b.** $0.07 + \frac{3}{100}$	**c.** $2.022 + \frac{33}{1000}$

4. Solve without using a calculator.

a. 2.31×0.04

b. $3.38758 \div 7 + 0.045$

5. Round to...

	0.0882717	**0.489932**	**1.299959**
the nearest thousandth			
the nearest hundred-thousandth			

6. Multiply or divide mentally.

a. $0.24 \div 3 =$	**b.** $5.4 \div 0.6 =$	**c.** $0.081 \div 0.009 =$
d. $2 \times 0.05 =$	**e.** $8 \times 0.009 =$	**f.** $11 \times 0.0005 =$

7. A rectangular plot of land has sides that measure 50.5 m and 27.6 m.
This plot is then divided into four equal pieces. What is the area of each fourth?

8. Multiply or divide these decimal numbers.

a. $1{,}000 \times 0.02 =$	**b.** $100 \times 0.0047 =$
c. $10^6 \times 1.097 =$	**d.** $0.6 \div 100 =$
e. $12.45 \div 10{,}000 =$	**f.** $324 \div 10^5 =$

9. Find the value of the expression $0.04 \div y$, when

a. $y = 4$	**b.** $y = 0.04$	**c.** $y = 10$

10. Change into the basic unit (meter, liter, or gram).

a. 56 mm

b. 9 km

c. 9 cg

d. 16 dl

11. Convert the measurements into the given units.

a. 2.7 km = __________ m = __________ cm = __________ mm

b. 5,600 ml = __________ cl = __________ dl = __________ l

c. 0.6 g = __________ dg = __________ cg = __________ mg

12. A newborn baby weighs 7 pounds 6 ounces.
Is this more or less than 7.4 pounds?

13. Which is a better deal:
A 1-pint bottle of honey that costs $7,
or a 24-oz bottle of honey that costs $12?

14. Divide, giving your answer as a decimal. If necessary, round the answers to three decimal digits.

a. $5.36 \div 0.2$	**b.** $1.6 \div 0.05$
c. $22.9 \div 7$	**d.** $\frac{8}{9}$

Mixed Review 3

1. Which power of ten is equal to a hundred million?

2. Write in expanded form using exponents.

   3,500,480

3. Estimate the result using mental math and rounded numbers. Find the exact value using a calculator. Also, find the error of estimation.

a. $213 \cdot 5{,}829$	**b.** $435{,}212 \div 993$
Estimation:	Estimation:
Exact value:	Exact value:
Error of estimation:	Error of estimation:

4. Evaluate the expression for the given values of the variable c.

c	$\mathbf{c + \frac{2c}{5}}$
10	$10 + \frac{2 \times 10}{5} = 10 + 4 = 14$
15	

c	$\mathbf{c + \frac{2c}{5}}$
20	
25	

5. The gas gauge shows 5.1 gallons of gasoline left, and that is 3/10 of the volume of the gas tank. How much does the gas tank hold when full?

6. Eric bought two printers. One cost $98 and the other cost 6/7 of that price. Find the total cost.

7. Simplify.

a. $\frac{15 + 150}{5}$	**b.** $\frac{5}{15 + 5}$	**c.** $\frac{380 + 10}{12 - 9}$

8. Write an expression.

a. The quantity $t - 1$ squared.

b. x less than 9.

c. 7 more than S.

d. 8 times the sum of 4, x, and 2.

e. The quotient of x^2 and the quantity $x + 1$.

9. Evaluate the expressions for the given value of the variable.

a. $3x - 11$ when $x = 8$	**b.** $\frac{3}{z} \cdot 7$ when $z = 5$

10. Simplify the expressions.

a. $x \cdot x \cdot x \cdot x \cdot x$	**b.** $p + 2 + p$
c. $5 \cdot x \cdot x \cdot 2 \cdot x$	**d.** $9z - 2z + z$
e. $f + f + x + x + f$	**f.** $6 + s + 2s + 4$

11. Write an inequality for each phrase. You will need to choose a variable to represent the quantity in question.

a. The AC runs at least 18 hours per day.

b. The jacket can cost a maximum of $40.

c. She is over 12 years old.

12. Solve the inequality $x + 1 < 8$ in the set {3, 4, 5, 6, 7, 8}.

13. Multiply using the distributive property.

a. $3(5x + 6) =$	**b.** $2(8x + 2 + y) =$

14. Solve the equations.

a. $x + 78 = 412$	**b.** $\frac{x}{9} = 600$	**c.** $y - 5 = 12 + 18$
$=$	$=$	$=$
$=$	$=$	$=$

Mixed Review 4

1. Four parents shared the cost of $207.48 for hosting a parent meeting in this way: one parent paid half of the cost, and the rest was divided equally between the rest of the parents. How much did each parent pay?
Hint: you can draw a bar model to help.

2. Write in normal form (as a number).

a. $3 \cdot 10^8 + 2 \cdot 10^7 + 9 \cdot 10^6 + 3 \cdot 10^2$

b. $1 \cdot 10^6 + 5 \cdot 10^4 + 3 \cdot 10^0$

3. Round to the place of the underlined digit. Be careful with the nines!

a. $5{,}69\underline{9}{,}528 \approx$ ________________ **b.** $219{,}99\underline{7}{,}101 \approx$ ________________

c. $8\underline{2}{,}788{,}000 \approx$ ________________ **d.** $3{,}999{,}9\underline{9}2{,}567 \approx$ ________________

4. Evaluate the expression for the given values of the variable x.

Variable	Expression $\frac{x^2}{3}$	Value
$x = 1$	$\frac{1^2}{3}$	$\frac{1}{3}$
$x = 2$		

Variable	Expression $\frac{x^2}{3}$	Value
$x = 3$		
$x = 5$		

5. Write an expression for each scenario, and then find the value of the expression.

a. The sum of 12 and 56 divided by 4.

b. The quotient of 8 and the quantity 4 to the third power.

6. Simplify the expressions.

a. $c \cdot c \cdot c \cdot 8 \cdot c$	**b.** $7c - 2c + 8$
c. $t + t + t + 3 - 2t$	**d.** $2x^2 + 5 + 11x^2 + 8$

7. Write an expression for each situation.

a. Anna has m marbles. She gave 1/3 of them to her friend.
How many marbles did her friend get?

b. Sadie is s years old. Fanny is 6 years younger than Sadie.
How old is Fanny?

c. How old will Sadie be in 5 years?

d. How old will Fanny be in 5 years?

8. Solve these equations.

a. $7x + 2x = 54$ $=$ $=$ $=$	**b.** $8r - 3r = 40$ $=$ $=$ $=$	**c.** $t \div 50 = 5 + 6$ $=$ $=$ $=$
d. $w - 88 = 20 \cdot 60$ $=$ $=$ $=$	**e.** $2x - 6 = 16$ $=$ $=$ $=$	**f.** $8x + 17 = 81$ $=$ $=$ $=$

9. Factor these sums thinking of the distributive property "backwards."

a. $16y + 12 =$ ____ (y + ____)	**b.** $9x + 9 =$ ____ (____ + ____)
c. $54c + 24 =$ ____ (____ + ____)	**d.** $15a + 45 =$ ____ (____ + ____)

10. Solve the equations.

a. $x + 4.5039 = 7$	**b.** $0.938208 - x = 0.047$	**c.** $2x = 6.0184$

Ratios Review

1. Write the equivalent ratios.

a. $\frac{4}{3} = \frac{20}{}$	**b.** 6 : 7 = 18 : _______	**c.** _______ to 30 = 2 to 15	**d.** $\frac{7}{3} = \frac{}{12}$

2. Simplify the ratios.

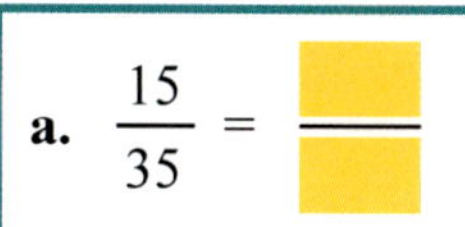

a. $\frac{15}{35} = \frac{}{}$	**b.** $\frac{6}{16} = \frac{}{}$	**c.** 33 : 30 = _____ : _____	**d.** 9 : 12 = _____ : _____

3. **a.** Draw a picture where there are 2 hearts for each 3 triangles and a total of 15 triangles.

b. Fill in the unit rates:

_______ hearts for 1 triangle

_______ triangles for 1 heart

4. A car traveled 300 miles in 6 hours at a constant speed. Fill in the table of equivalent rates:

Miles					
Hours	1	2	3	4	5

Miles					
Hours	6	7	8	9	10

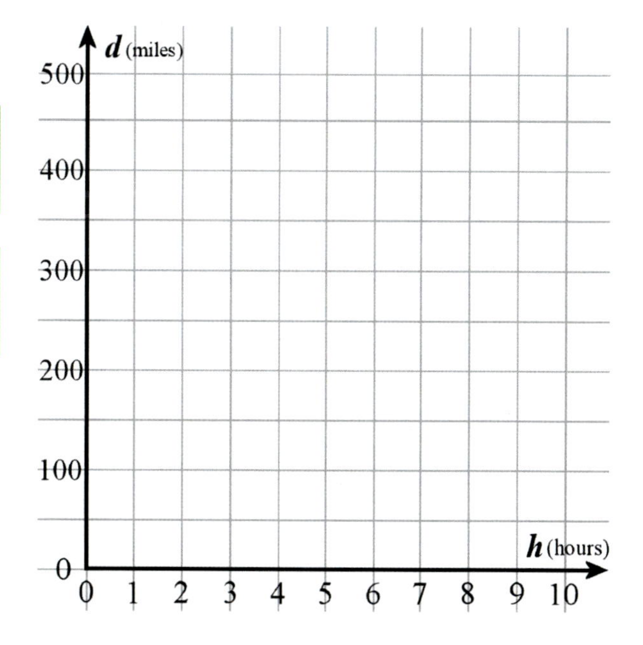

a. Plot the points in the coordinate grid.

b. What is the unit rate?

c. How far would the car go at that speed in 7 1/2 hours?

d. How long would it take for it to travel 225 miles?

5. A mixture of salt and water contains 20 grams of salt and 1200 grams of water.

a. Write the ratio of salt to water, and simplify it to lowest terms.

b. Use the same ratio of salt to water. If there are 100 grams of salt, how many grams of water would be needed?

6. Mom and Dad's ages are in a ratio of 11:12. Dad is 3 years older than Mom. How old are Mom and Dad?

7. A bean plant is 3/5 as tall as a tomato plant. The tomato plant is 20 cm taller than the bean plant.

a. What is the ratio of the bean plant's height to the tomato plant's height?

b. How tall is the bean plant? The tomato plant?

8. The aspect ratio of a television screen is 16:9 (width to height), and it is 63 cm high. What is its width?

9. **a.** If 12 kg of chicken feed costs $20, how much would 5 kg cost?

b. What is the unit rate? (price per 1 kg)

10. Use ratios to convert the measuring units. 1 kg = 2.2 lb, and 1 ft = 30.48 cm. Round to one decimal digit.

a. 134 lb into kilograms
b. 156 cm into feet

Ratios Test

1. Write the equivalent ratios.

a. $\frac{3}{5} = \frac{18}{\square}$	b. 2 : 3 = 18 : ____	c. ____ to 45 = 2 to 9	d. 12 : 30 = ___ to 5

2. **a.** Draw a picture where there are 4 rectangles for each 3 triangles, and a total of 16 rectangles.

   b. Fill in the unit rates:

   _______ rectangles for **1** triangle

   _______ triangles for **1** rectangle

3. Fill in the missing numbers to form equivalent rates.

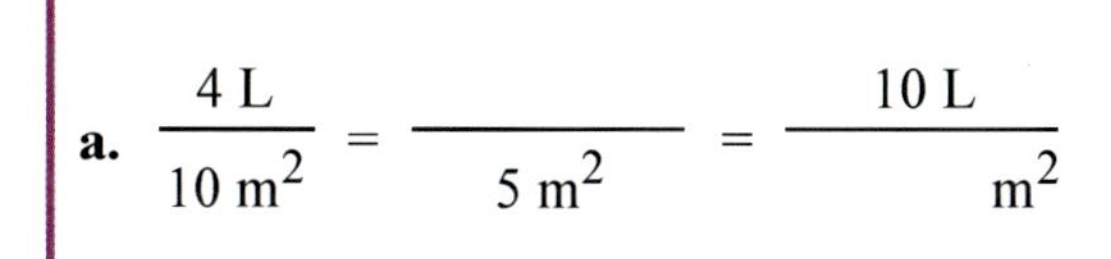

a. $\frac{4\text{ L}}{10\text{ m}^2} = \frac{\_\_\_\_}{5\text{ m}^2} = \frac{10\text{ L}}{\_\_\_\text{ m}^2}$	b. $\frac{\$9}{6\text{ min}} = \frac{\_\_\_\_}{2\text{ min}} = \frac{\_\_\_\_}{10\text{ min}} = \frac{\_\_\_\_}{1\text{ hour}}$

4. A mole can dig 3.6 meters in 36 minutes.

   a. What is the unit rate?

   b. Digging at the same speed, how far can the mole dig in 17 minutes?

5. You can buy 14 song downloads for $2.10.
   How much would 3 songs cost?

6. The length and width of a rectangle are in a ratio of 8:5.
   The shorter side is 15 cm.

   a. Find the longer side of the rectangle.

   b. Find the area of the rectangle.

7. You are mixing juice concentrate and water in a ratio of 1:7. How much water and how much concentrate do you need to make 4 liters of diluted juice?

8. A large passenger airplane burns about 35 gallons of fuel per 7 miles.

   a. Write a rate from this, and simplify it to the lowest terms.

   b. How far can the airplane travel with 500 gallons of fuel?

   c. How many gallons will the airplane need to travel 150 miles?

9. Anita and Michael divided a job of folding advertisements for inserts in 1,200 newspapers in a ratio of 3:5. Calculate how many inserts each one of them folded.

10. Use ratios to convert the measuring units. 1 in = 2.54 cm, and 1 ft = 30.48 cm.

a. 60 cm into inches
b. 4.5 feet into cm

Mixed Review 5

1. Write the division equation, if the calculation to check it is $13 \times 381 + 5 = 4{,}958$.

2. Solve $43 \div 9$ to three decimal digits.

3. Round to the place of the underlined digit. Be careful with the nines!

a. $51{,}99\underline{9},601 \approx$ ______________ **b.** $109{,}9\underline{9}9{,}339 \approx$ ______________

4. Multiply or divide mentally.

a. $3 \times 0.25 =$ ______	**b.** $8 \times 0.08 =$ ______	**c.** $1 \div 0.05 =$ ______	**d.** $0.99 \div 11 =$ ______
$4 \times 0.025 =$ ______	$100 \times 0.0008 =$ ______	$4 \div 0.05 =$ ______	$0.06 \div 0.001 =$ ______

5. Multiply or divide the decimals by the powers of ten.

a. $10^5 \times 3.07 =$	**b.** $10^4 \times 0.00078 =$
c. $12.7 \div 10^3 =$	**d.** $5{,}600 \div 10^5 =$

6. The area of a square is $4y^2$. What is the length of one side?

7. Fill in the table.

Expression	the terms in it	coefficient(s)	Constants
$2a + 3b$			
$10s$			
$11x + 5$			
$8x^2 + 9x + 10$			
$\frac{1}{6}p$			

8. Xavier and Yvonne got 10 small cookies from their mom to share. They did not have to share them equally. Let us consider the cookies Xavier got (X) and the cookies Yvonne got (Y).

   a. Fill in the table with possible values for X and Y, and plot the points in the grid.

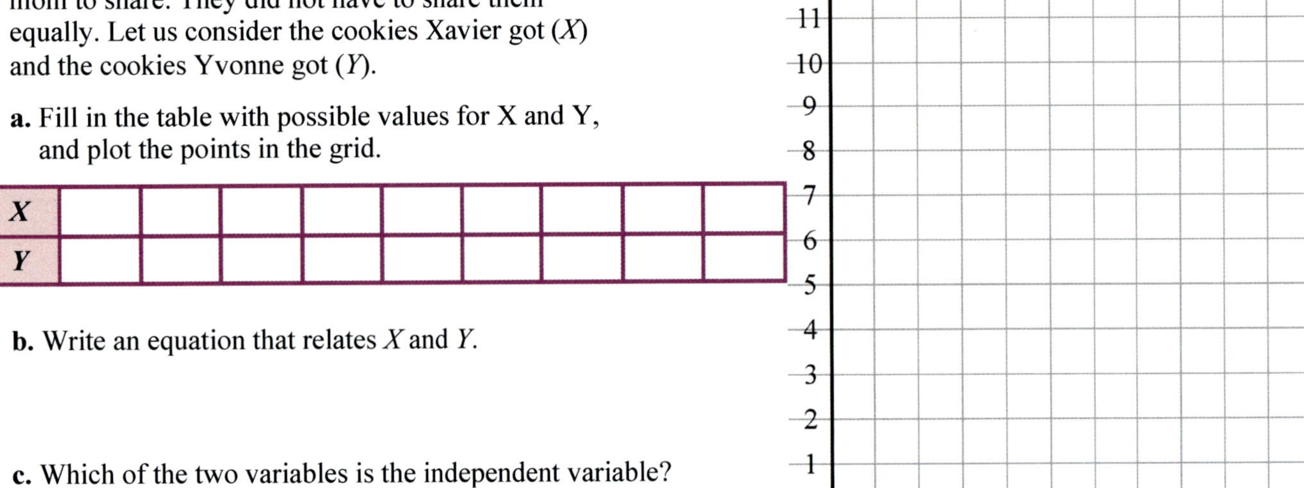

X									
Y									

   b. Write an equation that relates X and Y.

   c. Which of the two variables is the independent variable?

9. Write an expression for each situation.

   a. The value, in cents, of n nickels (n is a whole number).

   b. You have 67 drawings and you throw away y of them. How many do you have now?

   c. The original price of a puzzle is p. Now it is discounted and costs only 8/10 as much. What is the current price?

10. Divide, giving your answer as a decimal. If necessary, round the answers to three decimal digits.

a. $675.5 \div 0.3$	**b.** $\frac{2}{7}$

Mixed Review 6

1. Write each statement as an equation. Then solve the equation.

a. The quotient of a secret number and 11 is equal to 12.	**b.** The sum of 3, 5, and a certain number is 105.

2. Solve the equations.

a. $x \div 6 = 40 + 50$	**b.** $1{,}000 - x = 40 \cdot 6$	**c.** $8x + 2x = 15 \cdot 6$

3. The numbers below are prices for sets of 12 colored pencils from seven different stores.

$3.89 $3.99 $4.45 $3.79 $4.10 $4.19 $4.02

a. Find the average price.

b. How much will a teacher save if she buys 100 sets of the pencils at the cheapest price as compared to the most expensive price?

4. Calculate.

a.	**b.**	**c.**
$10 \cdot 0.009 =$ $0.5 \cdot 0.6 =$	$40 \cdot 0.08 =$ $1{,}000 \cdot 1.2 =$	$0.1 \cdot 0.2 \cdot 0.3 =$ $0.11 \cdot 0.02 =$
d.	**e.**	**f.**
$10 \div 0.2 =$ $0.6 \div 0.2 =$	$0.075 \div 0.025 =$ $0.3 \div 0.02 =$	$2.36 \div 2 =$ $0.0045 \div 5 =$

5. Write the amounts in basic units (meters, grams, or liters).

a. 6 kg = __________ g	**b.** 7 dam = __________ m	**c.** 7 kl = __________ L
5 dl = __________ L	5 hl = __________ L	50 mg = __________ g
5 mm = __________ m	30 cg = __________ g	8 cm = __________ m

6. We often compare the size of people and animals by comparing their weights.
Tim weighs 45 kg and a grasshopper weighs 3,000 mg.

a. How many times more does Tim weigh than the grasshopper?

b. Assuming that they were somehow packaged to be carried, could you carry the weight of a thousand grasshoppers?

7. Elaine paid 1/5 of her salary in taxes. After that, she paid 1/6 of what was left for rent.
Then she had $1,000 left. How much was her salary?

rent

tax

8. Divide. If necessary, round your answer to three decimal digits.

a. $45.7 \div 0.02$	**b.** $928 \div 0.003$	**c.** $\frac{5}{8}$

Percentage Review

1. Find a percentage of a number	2. A fractional part as a percentage
What is 60% of 300 miles? Calculate 0.6 × 300 miles = 180 miles. Or, using mental math, first calculate 10% of 300 miles, which is 1/10 of it, or 30 miles. Then multiply 6 × 30 miles = 180 miles. *Of the 15,400 workers in a city, 22% work in a steel factory. How many workers is that?* Calculate: 0.22 × 15,400 = 3,388 workers.	*What percentage is 600 g of 2 kg?* Write the fraction $\frac{600\text{ g}}{2{,}000\text{ g}} = \frac{6}{20} = \frac{30}{100} = 30\%$. *One backpack costs \$18 and another costs \$29. What percent is the price of the cheaper backpack of the price of the more expensive one?* Write the fraction $\frac{\$18}{\$29} = 0.6206... \approx 62\%$.
1. Change the percentage into a decimal. 2. Then multiply the number by that decimal. Alternatively, use mental math shortcuts for finding 5%, 10%, 20%, 25%, 50%, *etc.* of a number.	1. First write the fraction. Note that the two quantities in the fraction must both be in the same units: both grams, both meters, both dollars, *etc.* 2. Then convert the fraction into a decimal and finally a percentage.

1. Write as percentages, fractions, and decimals.

a. _____% = $\frac{68}{100}$ = _____	**b.** 7% = $\frac{\quad}{\quad}$ = _____	**c.** _____% = $\frac{\quad}{\quad}$ = 0.15
d. 120% = $\frac{\quad}{\quad}$ = _____	**e.** _____% = $\frac{224}{100}$ = _____	**f.** _____% = $\frac{\quad}{\quad}$ = 0.06

2. Fill in the table. Use mental math.

percentage ↓ number →	**6,100**	**90**	**57**	**6**
1% of the number				
4% of the number				
10% of the number				
30% of the number				

3. A skating group has 15 girls and 5 boys.
What percentage of the skaters are girls?

4. Write as percentages. You may need long division in some problems.
   If necessary, round your answers to the nearest percent.

   a. 3/4

   b. 2/25

   c. 1 5/8

5. Emma is 5 ft 4 in tall and Madison is 4 ft tall. How
   many percent is Emma's height of Madison's height?

6. A cheap chair costs $25. The price of another chair is 140% of that.
   How much does the other chair cost?

7. A bag has 25 green marbles and some white ones, too. The green marbles are 20% of the total.
   How many marbles are there in total? How many white marbles are there?

8. Andrew earns $2,000 monthly. He pays $540 of his salary in taxes.
   What percentage of his income does Andrew pay in taxes?

9. Which is cheaper, an $18 shirt discounted by 20%,
   or a $16 shirt discounted by 10% ?

10. (*Challenge*) One square has sides that are 2 cm long, and another has sides that are 4 cm long.
    How many percent is the area of the smaller square of the area of the larger square?

Percentage Test

A calculator is not allowed.

1. Write as percentages, fractions, and decimals.

a. ______% $= \frac{45}{100} =$ ______	**b.** 179% $= \frac{\quad}{\quad} =$ ______	**c.** ______% $= \frac{\quad}{\quad} = 0.02$

2. Fill in the table. Use mental math.

percentage / number	**5,200**	**80**	**9**
1% of the number			
3% of the number			
70% of the number			

3. Write 4/7 as a percentage. Round your answer to the nearest percent.

4. A toy costs $12. It is discounted by 30%. What is the new price?

5. A cap costs $7.00. Another cap costs 120% of the price of the first cap. How much does the second cap cost?

6. A store got a shipment of 120 T-shirts. Forty percent of them are white. How many T-shirts are *not* white?

7. A store window shows 2 red caps and 8 green caps. What percentage of the caps are red?

8. The chess club has 24 members, of which 8 are girls.
   What percentage of the members are boys?

9. Annie is 144 cm tall and Jessie is 160 cm tall.
   What percent of Jessie's height is Annie's height?

10. Which is cheaper, $35 jeans discounted by 10%,
    or $40 jeans discounted by 20%?

    How many dollars cheaper is it?

11. Andrew pays 20% of his salary in taxes. Andrew paid $400 in taxes.
    Find Andrew's salary in dollars.

12. A town has 2,100 senior citizens, which is 15% of the total population of the town.
    Calculate the total population of the town.

Mixed Review 7

1. Divide using long division in your notebook. Then, check your result.

a. $339{,}427 \div 26 =$ ________ R ________ ______ $\times$ ______ $+$ ______ $=$ ________	**b.** $6{,}594 \div 145 =$ ________ R ________ ______ $\times$ ______ $+$ ______ $=$ ________

2. Compare and write <, >, or = .

a. $659{,}000$ ☐ 10^6	**b.** 10 billion ☐ 10^9	**c.** $10^6 + 10^2$ ☐ $1{,}001{,}000$
d. 4^3 ☐ 3^4	**e.** 2^3 ☐ 3^2	**f.** 9×10^4 ☐ 2×10^5

3. Evaluate the expressions when the value of the variable is given.

a. $150 - 7s$ when $s = 9$	**b.** $\frac{3 + x}{x}$ when $x = 5$

4. Write in simplified form an expression for the area and an expression for the perimeter of the shape.

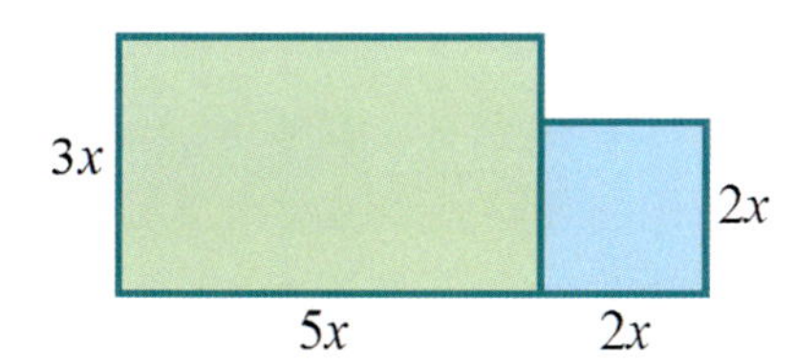

5. Simplify the expressions.

a. $y + 7 + 3y$	**b.** $r \cdot r \cdot r \cdot 8$

6. A typical ruler a student might use is 30 cm long.
How long would it be in inches?
(1 inch = 2.54 cm)

7. Choose the expressions that have the value 6.

a. $18 \div 3$ **b.** $1.8 \div 0.03$ **c.** $0.18 \div 0.03$ **d.** $1.2 \div 0.2$

e. $360 \div 6$ **f.** $3.6 \div 0.6$ **g.** $0.00036 \div 0.00006$ **h.** $0.9 \div 1.5$

i. $0.9 \div 0.15$ **j.** $0.009 \div 0.0015$ **k.** $0.0012 \div 0.002$ **l.** $0.0006 \div 0.0001$

8. One paper clip weighs 14 dg. They are sold in boxes of 200.

a. Calculate the weight of the box, in grams.

b. If someone wanted 1 kg of paperclips, how many boxes would he need to buy?

9. Sandra gets paid \$6 for every 15 minutes she works. Fill in the missing numbers to form equivalent rates.

$$\frac{\quad}{5\text{ min}} = \frac{\$6}{15\text{ min}} = \frac{\quad}{20\text{ min}} = \frac{\quad}{25\text{ min}} = \frac{\quad}{1\text{ hr}}$$

10. The width and length of a rectangle are in a ratio of 1:7, and its perimeter is 120 mm. Find the rectangle's width and length.

11. On average, Gary makes a basket eight times out of every ten shots. How many baskets can he expect to make when he practices 25 shots?

12. Solve the equations.

a. $312 = x + 78$	**b.** $\frac{z}{2} = 60 + 80$	**c.** $7y - 2y = 45$
=	=	=
=	=	=

13. The formula $m = 0.3048f$ can be used to convert feet into meters. The variable f is the length in feet, and the variable m is the length in meters. Use the formula to convert 89 feet into meters. Give your answer to two decimals.

14. A car travels with a steady speed of 24 miles per 30 minutes. Fill in the table.

Distance		24 miles				
Time	10 min	30 min	50 min	1 hour	2 1/2 hours	3 hours

Mixed Review 8

A calculator is allowed only in the last problem.

1. Write as decimals.

a. 392 hundred-thousandths

b. 5 and 15 ten-thousandths

c. 23 millionths

d. 12 and 12 thousandths

2. Write as fractions.

a. 0.000016

b. 2.9381

c. 0.39402

3. Find the value of the expression $y - 0.05$ when

a. $y = 1$	**b.** $y = 0.1$	**c.** $y = 1.1$

4. Round to...

	2.97167	**0.046394**	**2.33999**	**1.199593**
the nearest tenth				
the nearest thousandth				

5. Multiply both the dividend and the divisor by the same number, so that the divisor will be a whole number. Then divide mentally.

a. $\dfrac{5.6}{0.4} = \underline{\qquad} =$	**b.** $\dfrac{4}{0.02} = \underline{\qquad} =$	**c.** $\dfrac{0.9}{0.003} = \underline{\qquad} =$

6. When 1,200 people were polled about their favorite foods, 320 said they liked mashed potatoes best.

a. Write a ratio, and simplify it to the lowest terms.

b. Assuming the same ratio holds true in another group of 150 people, how many of those people can we expect to like mashed potatoes as their favorite food?

7. Fill in the missing numbers to form equivalent rates.

a. $\frac{14\text{ km}}{20\text{ min}} = \frac{\_\_\_\_\_\_}{5\text{ min}} = \frac{\_\_\_\_\_\_}{45\text{ min}}$	**b.** $\frac{\_\_\_\_\_\_}{8\text{ bottles}} = \frac{\_\_\_\_\_\_}{1\text{ bottle}} = \frac{\$42}{10\text{ bottles}}$

8. You need 2 kg of fertilizer for every 120 m^2 of lawn.
How much fertilizer would you need for a rectangular 15 m by 20 m lawn?

9. Calculate the values of y according to the equation $y = 2x - 4$.

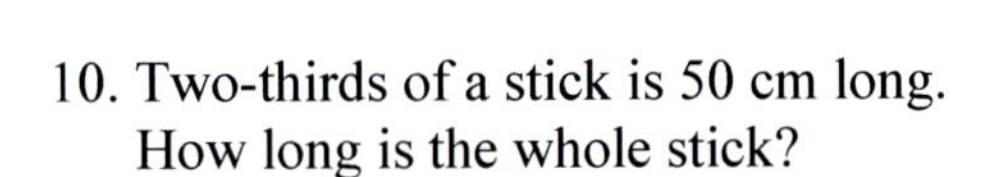

x	2	3	4	5	6	7
y						

Then plot the points.

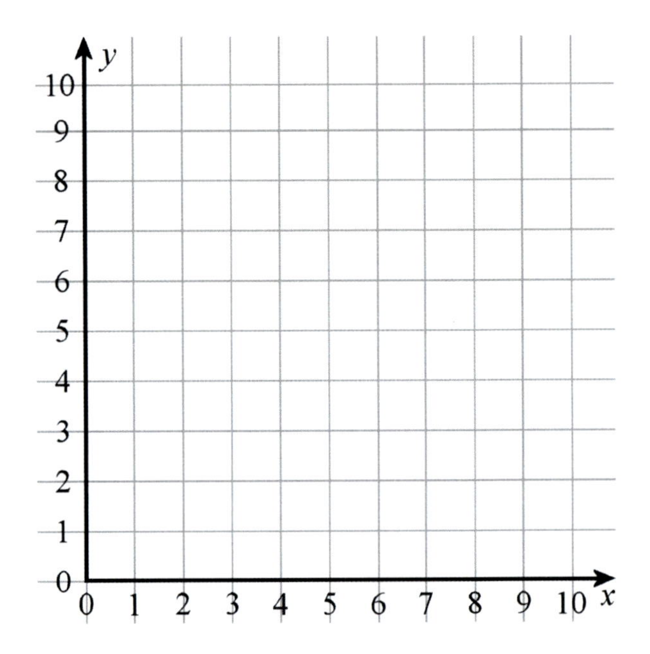

10. Two-thirds of a stick is 50 cm long.
How long is the whole stick?

11. Marsha has 2 gallons of punch, which she is pouring into 6-oz servings.
How many servings will she be able to get?

12. The children Hannah, 120 cm, and Erica, 1.05 m, stand on stools to see how tall they are.
At what height are the tops of their heads, if the girls stand on stools with the heights of:

a. 3.1 dm

b. 550 mm

c. 45 cm

13. Convert the given distances into metric units. Round the numbers to one decimal place. *Use a calculator* and the conversion factors at the right. →

1 inch = 2.54 cm
1 foot = 0.3048 m
1 mile = 1.6093 km

Every afternoon Erica bicycles 5 miles (_________ km) to the horse ranch.

Erica takes care of a horse that is 15 *hands*, or 60 inches (_____________ m), tall.

She likes to go riding on a trail that is 4 mi 500 ft (_________ km) long.

Prime Factorization, GCF, and LCM Review

1. Factor the following numbers into their prime factors.

a. 81 / \	**b.** 26 / \	**c.** 65 / \
d. 96 / \	**e.** 124 / \	**f.** 450 / \

2. Simplify.

a. $\frac{28}{84} = \frac{4 \times 7}{21 \times 4} =$	**b.** $\frac{75}{160} =$
c. $\frac{222}{36} =$	**d.** $\frac{48}{120} =$

3. Find the least common multiple of these pairs of numbers.

a. 3 and 7	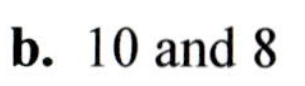 **b.** 10 and 8
c. 11 and 6	**d.** 6 and 8

4. Find the greatest common factor of the given number pairs.

a. 24 and 64	**b.** 100 and 75
c. 80 and 96	**d.** 78 and 96

5. Fill in with the words "multiple(s)" or "factor(s)."

- 25, 50, 75, 100, 125, and 150 are ____________________ of 25.
- 1, 2, 5, 10, 25, and 50 are ______________________ of 50.
- Each number has an infinite number of __________________________.
- Each number has a greatest ________________.
- If a number x divides into another number y, we say x is a ____________________ of y.

b. List five different multiples of 15 that are less than 200 but more than 60.

c. Find five numbers that are multiples of both 4 and 7.
What is the LCM of 4 and 7?

6. First, find the GCF of the numbers. Then factor the expressions using the GCF.

a. GCF of 12 and 21 is ______ 12 + 21 = ____ · ____ + ____ · ____ = ____(____ + ____)
b. GCF of 45 and 70 is ______ 45 + 70 = ____(____ + ____)

7. Draw two rectangles, side by side, so that they represent the sum 42 + 30 and share one side.

Prime Factorization, GCF, and LCM Test

1. Factor the following numbers into their prime factors.

a. 56	**b.** 90	**c.** 101
/ \	/ \	/ \

2. Find the least common multiple of these pairs of numbers.

a. 8 and 6	**b.** 6 and 12

3. Find the greatest common factor of the given number pairs.

a. 98 and 100	**b.** 98 and 35

4. Find four numbers that are multiples of both 6 and 10.

5. Find the LCM of 8 and 10 and the GCF of 8 and 10, and multiply them. What is the product?

6. Which number is a factor of all numbers?

7. Choose two primes between 10 and 30. What is their greatest common factor?

8. First, find the GCF of the numbers. Then factor the expressions using the GCF.

a. The GCF of 24 and 30 is ______ 24 + 30 = ____ · ____ + ____ · ____ = ____(____ + ____)
b. The GCF of 22 and 121 is ______ 22 + 121 = ____(____ + ____)

9. Simplify.

a. $\frac{124}{72} =$	**b.** $\frac{65}{105} =$

Mixed Review 9

1. Solve.

a. $10^4 \cdot 3$	**b.** 7^3	**c.** $10 \cdot 5^3$

2. Write in expanded form using exponents.

a. 109,200

b. 7,002,050

3. Andrew cut a 9-foot board into two pieces that are in a ratio of 3:5.
Find the length of each of the two pieces.

4. Convert each division problem into another, equivalent division problem that you can solve in your head.

a. $\dfrac{16}{0.4} = \dfrac{\quad}{\quad} =$	**b.** $\dfrac{7}{0.007} = \dfrac{\quad}{\quad} =$	**c.** $\dfrac{99}{0.11} = \dfrac{\quad}{\quad} =$

5. Multiply.

a. $100 \times 0.2 =$ ________	**b.** $3 \times 1.02 =$ ________	**c.** $0.9 \times 0.2 \times 0.5 =$ ________
$120 \times 0.02 =$ ________	$5 \times 3.02 =$ ________	$30 \times 0.005 \times 0.2 =$ ________

6. **a.** Draw a bar model to represent this situation:
The ratio of girls to boys in a vocational school is 7:4.

b. What is the ratio of boys to all students?

c. If there are 748 students in all, how many are girls?
How many are boys?

7. Liz is 150 cm tall, and her dad is 1.8 m tall. What percentage is Liz's height of her dad's height?

8. The Madison family spent $540 for groceries in one month. That was 24% of their total budget. How much was their total budget for the month?

9. Which is cheaper, a $180 camera discounted by 20%, or a $155 camera discounted by 10%?

10. Multiply using the distributive property.

a. $2(7m + 4) =$	**b.** $10(x + 6 + 2y) =$

11. Write an expression.

a. The quantity $5s$ plus 8, divided by 7.

b. The quantity n plus 11, cubed.

c. y more than 8.

d. x divided by y squared.

12. Solve the inequality $x - 3 < 0$ in the set $\{-2, -1, 0, 1, 2, 3\}$.

13. Divide, giving your answer as a decimal. If necessary, round the answers to three decimal digits.

a. $0.928 \div 0.3$	**b.** $\frac{7}{34}$

Mixed Review 10

A calculator is allowed only in the last problem.

1. Write as percentages. If necessary, round your answers to the nearest percent.

   a. 4/5

   b. 17/20

   c. 5/11

2. Write the fractions from the previous problem as decimals.

   a. 4/5 =   **b.** 17/20 =   **c.** 5/11 =

3. A store got a shipment of 155 calculators.
   The ratio of basic calculators to scientific calculators was 4:1.

   a. Draw a bar model to represent the situation.

   b. What fractional part of the calculators were basic calculators?

   c. What percentage of the calculators were basic calculators?

   d. How many scientific calculators were there?

4. Mike rides his bike at a constant speed of 20 km/h. Fill in the table.

Distance				16 km	20 km	24 km			
Time	6 min	12 min	15 min		1 hour		2 hours	3 hours	3 1/2 hours

5. One flash drive costs $25 and another costs 15% more.
   Find the total cost of buying both.

6. Factor these sums, thinking of the distributive property "backwards."

a. $32t + 8 =$ ____ (____ + ____)	**b.** $8 + 12x =$ ____ (____ + ____)
c. $15y + 45 =$ ____ (____ + ____)	**d.** $35 + 42w =$ ____ (____ + ____)

7. Grace got 35 points out of 40 in a test. Convert her test score into a percentage.

8. Jack gave 4/5 of his 90 toy cars to his cousins.
Then he divided the rest of his cars equally with his brother.
How many cars does Jack have now?

9. Multiply or divide the decimals by the powers of ten.

a. $10^4 \times 0.092 =$	**b.** $1{,}000 \times 0.0004 =$
c. $456.29 \div 1{,}000 =$	**d.** $63 \div 10^5 =$

10. Change into the basic unit (meter, liter, or gram).

a. 1534 cm **b.** 334 ml **c.** 0.9 kg

11. Write an equation for each situation EVEN IF you could easily solve the problem without an equation! Then solve the equation.

a. A camera costs $85 more than a camera bag.
If the camera costs $162, how much does the camera bag cost?

b. Jennifer purchased a set of 8 towels for $52.
How much did one towel cost?

12. If three shirts cost $14.10, then how much do seven shirts cost?

13. Convert into the given units. Round your answers to 2 decimals if needed.

a. 79 oz = _____ lb _____ oz	**c.** 7.82 qt = __________ gal	**e.** 2.54 lb = __________ oz
b. 4 ft 11 in = __________ in	**d.** 0.265 mi = __________ yd	**f.** 6.8 ft = _____ ft _____ in

Fractions Review

1. Add.

a. $\frac{5}{12} + \frac{1}{3}$	b. $\frac{5}{7} + \frac{1}{6}$	c. $1\frac{3}{5} + \frac{7}{8}$

2. Subtract. First write equivalent fractions with the same denominator.

a. $6\frac{2}{3} \rightarrow$ $-\ 2\frac{1}{6} \rightarrow -$	b. $7\frac{1}{6} \rightarrow$ $-\ 2\frac{3}{5} \rightarrow -$	c. $8\frac{9}{11} \rightarrow$ $-\ 4\frac{1}{3} \rightarrow -$

3. The pictures show how much pizza is left. Find the given part of it. Write a multiplication sentence.

a. Find $\frac{3}{4}$ of	b. Find $\frac{1}{5}$ of 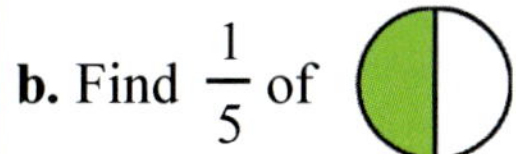 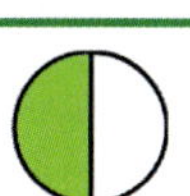	c. Find $\frac{2}{3}$ of 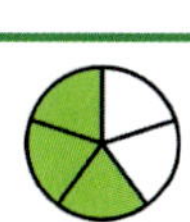

4. Multiply. Shade the areas to illustrate the multiplication.

a. 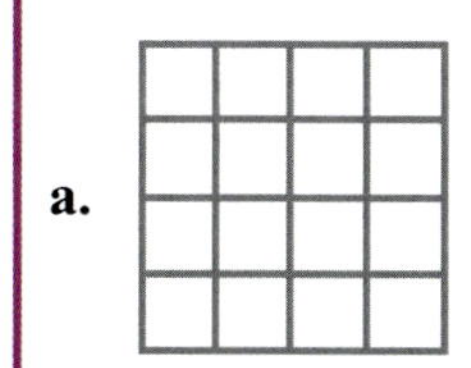 $\frac{1}{4} \times \frac{3}{4} =$	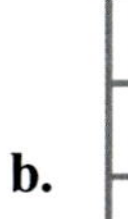 b. 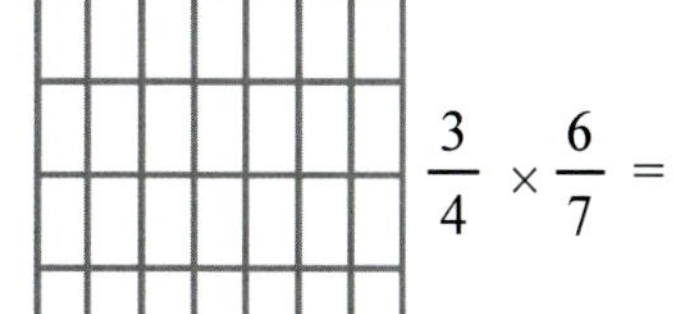 $\frac{3}{4} \times \frac{6}{7} =$

5. Simplify before you multiply.

a. $\frac{9}{12} \times \frac{6}{15}$

b. $\frac{3}{20} \times \frac{4}{21}$

c. $\frac{14}{40} \times \frac{10}{42}$

6. Write a division sentence for each problem and solve.

a. How many times does go into ?	**b.** How many times does go into 

7. Fill in the blanks and give an example. You can choose *any* number to divide by 4.

Dividing a number by 4 is the same as multiplying it by ____. Example:

8. Solve.

a. $\frac{2}{3} \div \frac{1}{5}$	**b.** $2\frac{1}{7} \div 1\frac{1}{2}$	**c.** $6 \div 1\frac{2}{3}$

9. A small, rectangular garden plot measures 7 1/2 feet by 4 3/8 feet.

a. Find its area.

b. Find its perimeter.

10. Write a real-life situation to match this fraction division: $\frac{9}{12} \div 3 = \frac{3}{12}$

11. How many 4 1/4 inch-pieces can you cut out of a 10-foot piece of string?

12. A 15-inch stick was cut into two pieces that were in the ratio of 1:7.
How long is each piece?

13. A model airplane is built to a scale of 1:15 compared to the real airplane. This means that the lengths, widths, and other measurements of the real airplane are 15 times as big as the corresponding measurements in the model. If the wingspan of the model is 32 1/4 in, what is the wingspan of the real airplane? Give your answer in feet and inches. *(Hint: You can multiply the whole-number part and the fractional part separately.)*

14. Five-sixths of the class went outside for recess, and 6 students stayed in the classroom.
How many students are in the whole class?

15. Two-fifths of a certain number is 160. What is the number?

16. Two farmers divided a day's kiwi fruit harvest. One farmer got 2/5 of the harvest and the other farmer got the rest. The farmer who got the least, gave 1/3 of his kiwi to his son, and kept 22 pounds.
How many pounds was the day's kiwi fruit harvest?

Puzzle Corner

a. Solve this "long" division!

$$\frac{1}{2} \div 5 \div 4 \div 3 \div 2 =$$

b. What did this division start with?

$$\frac{\square}{\square} \div 3 \div 5 \div 7 \div 9 = \frac{1}{1260}$$

Fractions Test

A calculator is <u>not</u> allowed.

1. Add or subtract

a. $\frac{5}{12} + \frac{1}{2} + \frac{5}{6}$	**b.** $\frac{5}{9} - \frac{2}{7}$
c. $2\frac{3}{10} + 2\frac{11}{12}$	**d.** $7\frac{1}{5} - 5\frac{7}{15}$

2. The Williams family had 3/4 of a pizza left over. The next day, Joe ate 3/4 of what was left. What part of the original pizza is left now?

3. How many 1/4-kg servings can you get from 5 1/3 kg of meat?

4. Joe had a board that is 4 1/3 ft long. He cut it into three equal pieces. How long are the pieces, in feet?

5. Multiply. Shade a rectangular area to illustrate the multiplication.

a. 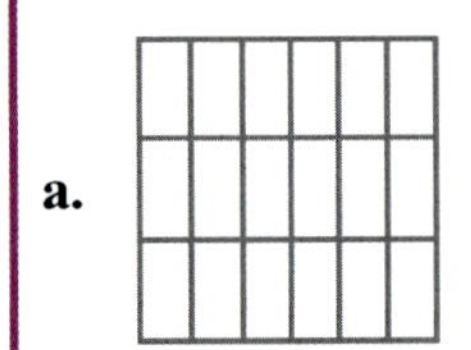 $\frac{2}{6} \times \frac{2}{3} =$

b. 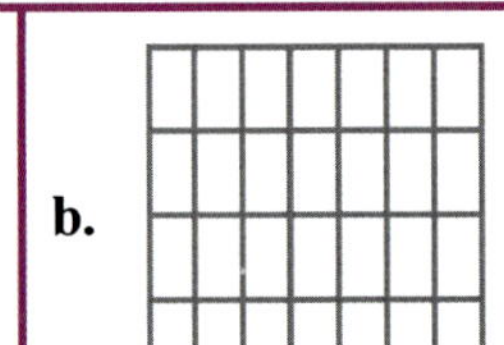 $\frac{3}{4} \times \frac{5}{7} =$

6. Solve.

a. $\frac{6}{7} \div \frac{1}{5}$	**b.** $\frac{12}{13} \div 2\frac{1}{3}$

7. A room measures 11 feet by 8 3/4 feet, and carpeting it costs $2.80 per square foot. Calculate the cost of carpeting the room.

8. Write a real-life situation to match this fraction division. Also, solve it. $2\frac{1}{2} \div 3 = ?$

9. How many 1 3/4 ft-pieces can you cut out of a 12-foot piece of string?

10. Mason and Aiden divided a $120 reward in a ratio of 2:3. Then, Aiden gave 3/10 of his money to his dad. How much does Aiden have now?

Mixed Review 11

1. **a.** One mile is 5,280 feet. *Estimate* how many inches are in one mile.

   b. Now *calculate* exactly how many inches are in one mile.

2. Jane mixed 2 parts of concentrated juice with 6 parts of water to make a total of 64 ounces of juice. How many ounces of concentrate and how many ounces of water were in the juice?

3. Write the equivalent rates.

a. $\frac{\$80}{4\text{ hr}} = \frac{\quad}{1\text{ hr}} = \frac{\quad}{3\text{ hr}} = \frac{\quad}{15\text{ min}}$	**b.** $\frac{2\text{ m}^2}{5\text{ min}} = \frac{10\text{ m}^2}{\quad} = \frac{\quad}{5\text{ hours}} = \frac{250\text{ m}^2}{\quad}$

4. A mixture of salt and water weighs 1.2 kg. It contains 2% salt by weight, and the rest is water. How many grams of salt and how many grams of water are in the mixture?

5. A train traveled 165 miles from one town to the next at an average speed of 90 mph. When did the train leave, if it arrived at 1440 hours (2:40 P.M.)?

6. Multiply or divide the decimals by the powers of ten.

a.	**b.**	**c.**
$10 \times 0.3909 =$ $1{,}000 \times 4.507 =$	$1.08 \times 100 =$ $0.0034 \times 10^4 =$	$10^6 \times 8.02 =$ $10^5 \times 0.004726 =$
d.	**e.**	**f.**
$0.93 \div 100 =$ $48 \div 10 =$	$3.04 \div 1{,}000 =$ $450 \div 10^4 =$	$98.203 \div 10^5 =$ $493.2 \div 10^6 =$

7. Factor the following numbers into their prime factors.

a. 65 /\	**b.** 75 /\	**c.** 82 /\

8. Find a number between 640 and 660 that is divisible by 3 and 7.

9. First, find the GCF of the numbers. Then factor the expressions using the GCF.

a. GCF of 16 and 42 is ______	**b.** GCF of 98 and 35 is ______
16 + 42 = ____ (____ + ____)	98 + 35 = ____ (____ + ____)

10. Draw two rectangles, side by side, to represent the sum 18 + 30.

11. Calculate the values of y according to the equation $y = 2x - 5$.

x	3	4	5	6	7	8
y						

Now, plot the points.

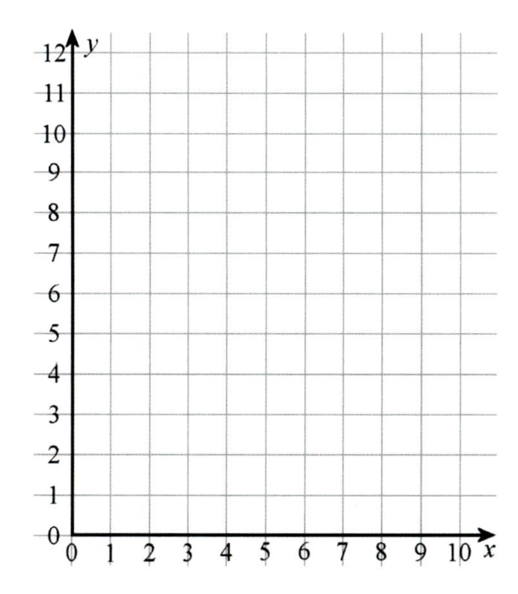

Mixed Review 12

A calculator is not allowed.

1. Find the least common multiple of the given number pairs.

a. 2 and 8	**b.** 6 and 9

2. Find the greatest common factor of the given number pairs.

a. 14 and 15	**b.** 48 and 60

3. Draw two rectangles to represent the sum 45 + 27 so that they share a side.

4. There is two quarts of ice cream in a container.
   This is divided equally between 9 people.
   How much will each person get (in ounces)?

5. Fill in.

   a. 11^2 gives us the ____________ of a ______________ with a side length of ______ units.

   b. 3×5^2 gives us the ____________ of _____ ______________ with a side length of ______ units.

   c. 4×0.4^3 gives us the ____________ of _____ ______________ with an edge length of ______ units.

6. Write in normal form (as a number).

a. $2 \times 10^7 + 6 \times 10^6 + 2 \times 10^4$	**b.** $1 \times 10^9 + 2 \times 10^5 + 8 \times 10^2 + 7 \times 10^0$

7. Factor the following numbers into their prime factors.

a. 99	**b.** 112	**c.** 200
/\	/\	/\

8. Chris has two kinds of containers for gasoline. The larger ones hold 8.5 liters, and the smaller ones hold 60% of that amount. What is the total capacity of three large and four smaller containers?

9. Samantha and George got paid $100 for working together on a project. Since Samantha had worked 5 hours and George only 3 hours, they decided it would be fair to divide the pay in a ratio of 5:3. How much more did Samantha earn than George?

10. Write an expression.

a. The quotient of 6 and $7s$.

b. Subtract $2x$ from 11.

c. The sum of x and 2, squared.

d. The quantity $5m$, cubed.

e. $2t^2$ divided by the difference of s and 1.

f. y less than 18.

11. Write an equation to match the bar model, and solve it.

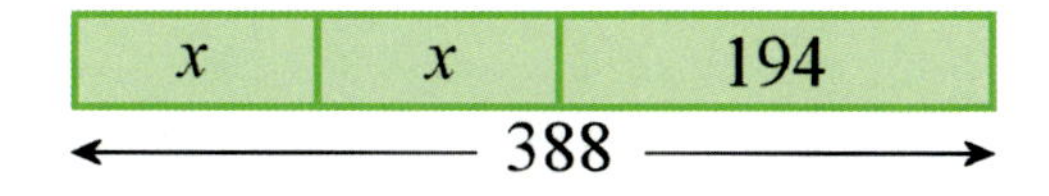

12. Multiply.

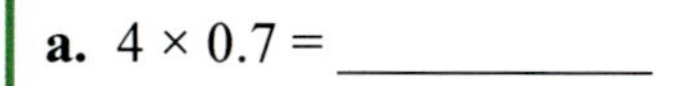

a. $4 \times 0.7 =$ ________	**c.** $3 \times 1.06 =$ ________	**e.** $10^5 \times 0.08 =$ ________
b. $50 \times 0.003 =$ ________	**d.** $100 \times 0.009 =$ ________	**f.** $40 \times 0.004 =$ ________

Integers Review

1. Compare. Write < or > in the box.

a. −1 ☐ −7	**b.** 2 ☐ −2	**c.** −6 ☐ 0	**d.** 8 ☐ −3	**e.** −8 ☐ −3

2. Order the numbers from the least to the greatest.

a. −6 2 −2 0	**b.** −14 −8 −11 −7

3. Express the situations using integers. Then compare them writing > or < in the box.

a. Lillian owes $12 and Hayley owes $18.	______ ☐ ______
b. At 2 PM, the temperature was 5°C below zero. Now it is 2°C.	______ ☐ ______
c. Joe rose in an elevator to the height of 16 m, whereas Gabriel went down 6 m below the ground.	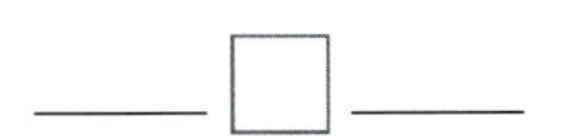

4. Simplify. In (e), write using a number.

a. | −11 | **b.** | 2 | **c.** | 0 | **d.** −(−19) **e.** the opposite of 7

5. Draw a number line jump for each addition or subtraction sentence.

a. −9 + 6 = ______ **b.** −2 + 5 = ______

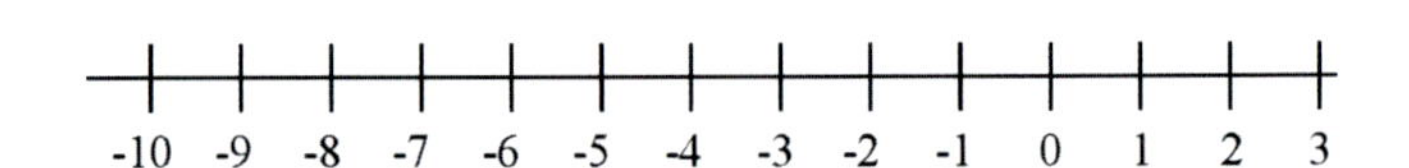

c. −3 − 5 = ______ **d.** 2 − 8 = ______

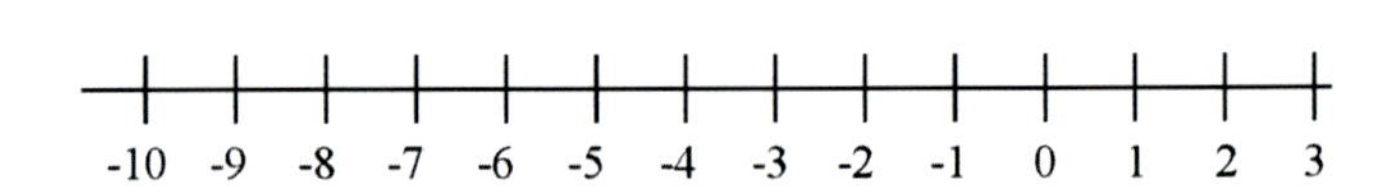

6. Write an addition or subtraction sentence.

a. You are at $^{-}10$. You jump 6 to the right. You end up at ______.

b. You are at $^{-}5$. You jump 8 to the right. You end up at ______.

c. You are at 3. You jump 7 to the left. You end up at ______.

d. You are at $^{-}11$. You jump 3 to the left. You end up at ______.

7. Add or subtract.

a.	**b.**	**c.**	**d.**
$2 + (-8) =$ ________	$-2 + (-9) =$ ________	$1 + (-7) =$ ________	$5 - (-2) =$ ________
$(-2) + 8 =$ ________	$2 - 8 =$ ________	$-4 - 5 =$ ________	$-3 - (-4) =$ ________

8. Write an addition or a subtraction sentence to match the situations.

a. May has \$35. She wants to purchase a guitar for \$85.
That would make her money situation be ____________.

b. A fish was swimming at the depth of 6 ft. Then he sank 2 ft.
Then he sank 4 ft more. Now he is at the depth of _________ ft.

c. Elijah owed his dad \$20. Then he borrowed another \$10.
Now his balance is _________.

d. The temperature was −13°C and then it rose 5°.
Now the temperature is ________ °C.

9. Use mathematical symbols to express these ideas

a. the distance of −17 from zero

b. the opposite of −11

10. Which expression below matches with the situation? Jacob owes more than fifty dollars.

a. balance > −\$50 **b.** balance = −\$50 **c.** balance < \$50 **d.** balance < −\$50

11. Plot the points from the function $y = 4 - x$ for the values of x listed in the table.

x	−5	−4	−3	−2	−1	0	1	2
y								

x	3	4	5	6	7	8	9
y							

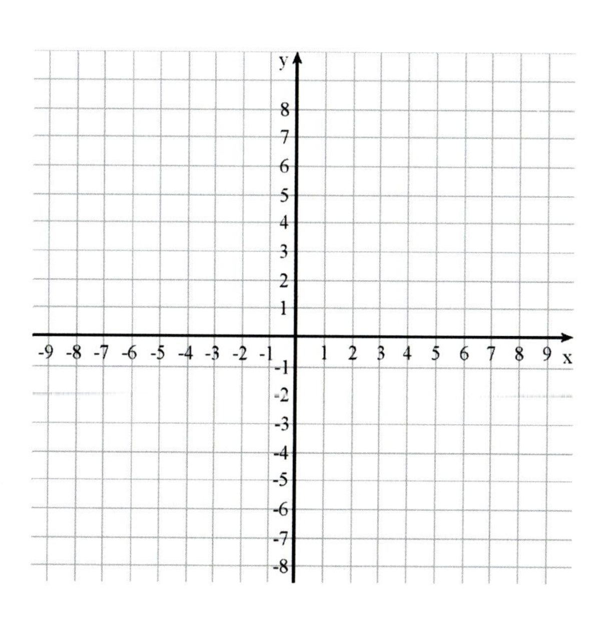

12. Find the missing integers.

a. $-2 + \_\_\_\_\_\_ = -8$	**b.** $4 + \_\_\_\_\_\_ = 0$	**c.** $5 - \_\_\_\_\_\_ = -2$
$3 + \_\_\_\_\_ = -2$	$-6 - \_\_\_\_\_\_ = -12$	$3 + \_\_\_\_\_\_ = 1$

13. Andrew drew a polygon, and then he reflected it in the x-axis. The vertices of the reflected polygon are: $(-9, 6)$, $(-6, 6)$, $(-9, 3)$, and $(-3, 0)$. What were the coordinates of the original vertices?

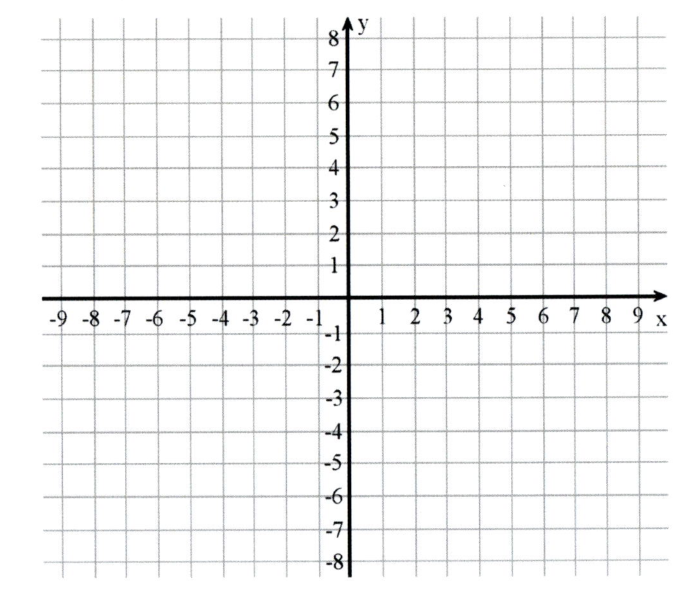

14. Find the distance between the points.

a. $(-3, -12)$ and $(-3, 15)$

b. $(-15, 34)$ and $(-21, 34)$

15. **a.** The points $(-7, -3)$, $(-1, -7)$, $(-1, -1)$, and $(-4, -6)$ are vertices of a quadrilateral. Draw the quadrilateral.

b. Reflect the quadrilateral in the y-axis. (Draw the new quadrilateral). Write the coordinates of the moved vertices.

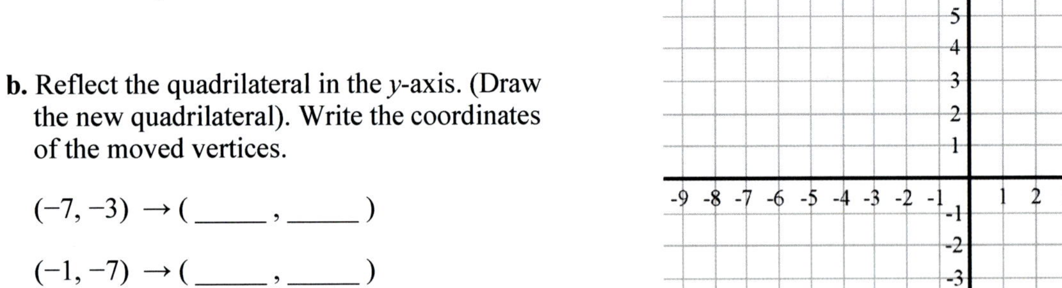

$(-7, -3) \rightarrow (\_\_\_\_\_, \_\_\_\_\_)$

$(-1, -7) \rightarrow (\_\_\_\_\_, \_\_\_\_\_)$

$(-1, -1) \rightarrow (\_\_\_\_\_, \_\_\_\_\_)$

$(-4, -6) \rightarrow (\_\_\_\_\_, \_\_\_\_\_)$

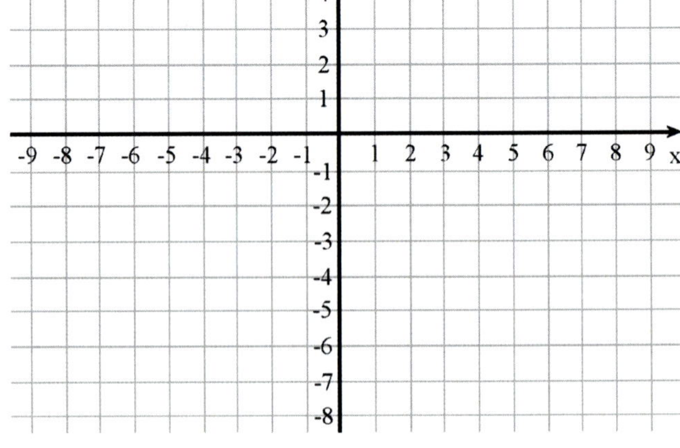

c. Now move the already reflected quadrilateral 7 units up. (Draw the new quadrilateral). Write the coordinates of the new vertices.

$(\_\_\_\_\_, \_\_\_\_\_)$ $(\_\_\_\_\_, \_\_\_\_\_)$ $(\_\_\_\_\_, \_\_\_\_\_)$ $(\_\_\_\_\_, \_\_\_\_\_)$

Integers Test

A calculator <u>is not</u> allowed.

1. Order the numbers 3, −3, −5, and 0 from the smallest to the greatest.

2. Draw a number line jump for each addition or subtraction sentence.

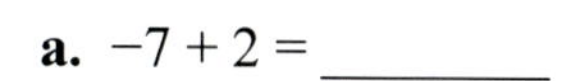

a. $-7 + 2 =$ ________

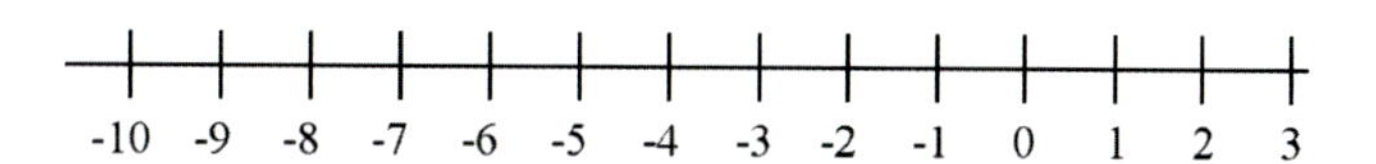

b. $-3 + 6 =$ ________

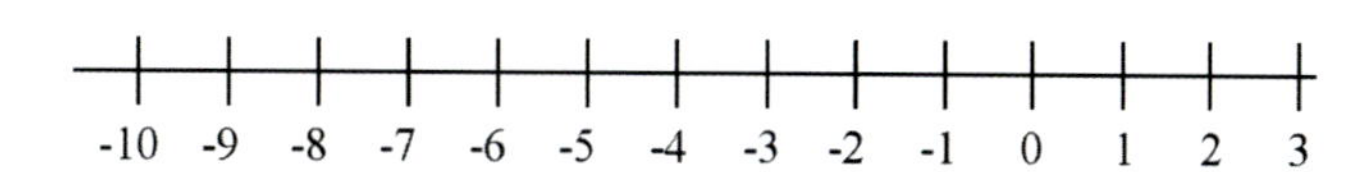

c. $-1 - 5 =$ ________

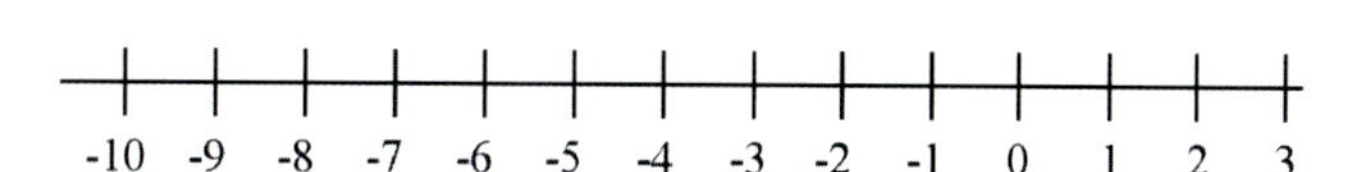

d. $2 - 7 =$ ________

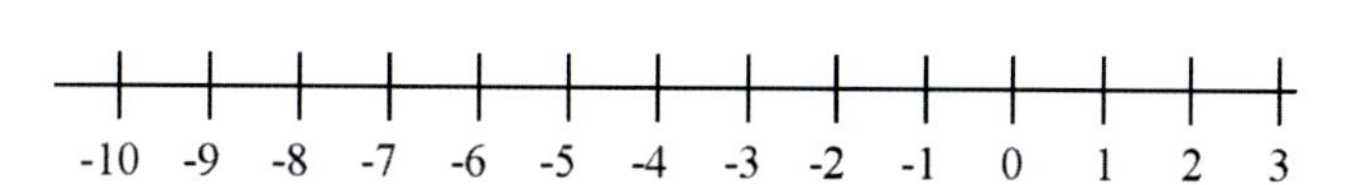

3. Add or subtract.

a.	b.	c.	d.
$3 + (-7) =$ ______	$(-1) + (-9) =$ ______	$4 + (-5) =$ ______	$8 - (-2) =$ ______
$(-3) + 7 =$ ______	$1 - 9 =$ ______	$-4 - 5 =$ ______	$-8 - (-2) =$ ______

4. Use mathematical symbols to express these ideas

a. The distance of −9 from zero.

b. The opposite of 43.

c. Henry's balance?
He owes some money, less than $20.

d. The temperature is colder than −10.

5. Write an addition or subtraction sentence to match the situations.

a. May owed $3. She borrowed $8 more.
Now her money situation is __________.

b. The temperature was 1°C and fell 4°.
Now the temperature is ________ °C.

c. A submarine was at the depth of 12 m. Then it rose 5 m.
It sank 10 m more. Now it is at the depth of ______ m.

6. Plot the points from the equation $y = x - 1$ for the values of x listed in the table.

x	−7	−6	−5	−4	−3	−2	−1	0
y								

x	1	2	3	4	5	6	7	8
y								

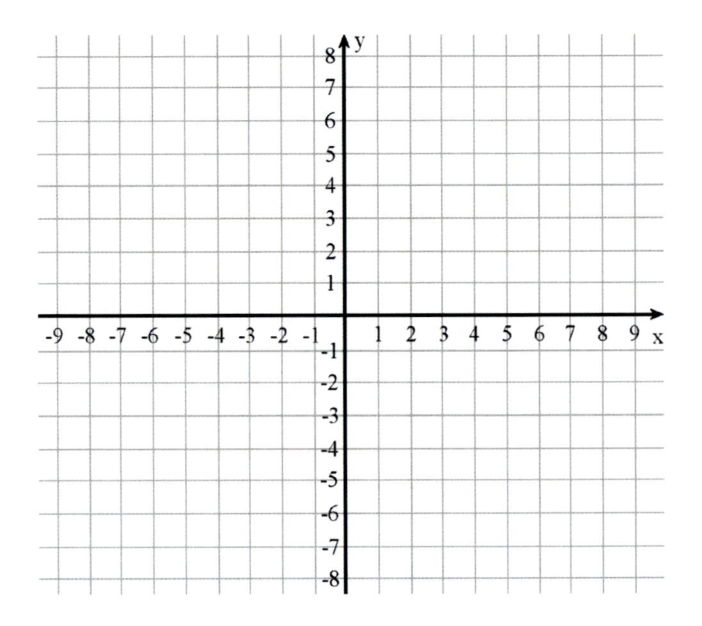

7. The points $(-7, -6)$, $(-5, -2)$, and $(-1, -4)$ are vertices of a triangle.

   a. Draw the triangle.

   b. Reflect it in the x-axis. Then move the already reflected triangle 5 units to the right.

   c. Draw the new triangle and write the coordinates of the new vertices.

   (_____, _____)

   (_____, _____)

   (_____, _____)

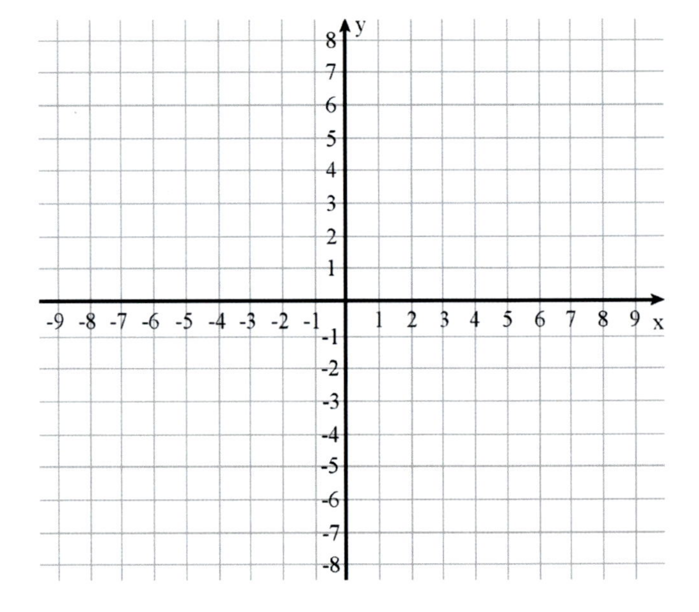

Mixed Review 13

1. Write an equation for each situation even if you could easily solve the problem without an equation! Then solve the equation.

   a. Katie is 54 years old. Shelly is 12 years younger than Katie. How old is Shelly?

   b. Bob bought some tulips for his wife. One tulip cost $2.15 The total cost was $45.15. How many tulips did Bob buy?

2. Find a number between 500 and 520 whose prime factorization has only 2s.

3. Write either a fraction multiplication or a division, and solve.

a. How many times does go into ?	**b.** How many times does go into ?
c. Find $\frac{3}{4}$ of	**d.** Find $\frac{2}{9}$ of

4. A package of cheap dominoes weighs 3 oz. A package of quality dominoes weighs 1 lb.

   a. How much does a box containing 50 packages of the cheap dominoes weigh? Give your answer in pounds and ounces.

   b. Another box contains 24 packages of the quality dominoes. Find how much more the box with quality dominoes weighs than the box with cheap dominoes.

5. Divide.

a. $2\frac{7}{8} \div \frac{2}{5}$	**b.** $4 \div 1\frac{5}{6}$
c. $5 \div \frac{2}{7}$	**d.** $10\frac{1}{10} \div \frac{3}{4}$

6. Annabelle can type 70 words in two minutes.
How many words can she type in 15 minutes?

7. Mom and Dad's ages are in the ratio of 7:8.
Dad is six years older than Mom. How old is Mom?

8. A rectangle's aspect ratio is 5:2, and
its perimeter is 84 cm. Find its area.

9. Keith paid $414 of his salary in taxes. After that, he had $1,459 left.
What percentage of his income did Keith pay in taxes?

10. Express these rates in the lowest terms.

a. 720 km : 4 hr	**b.** 6 kg for $4.20	**b.** 120 miles on 5 gallons

11. Simplify before you multiply.

a. $\frac{5}{36} \times \frac{24}{45}$	**b.** $\frac{16}{30} \times \frac{25}{24}$	**c.** $\frac{14}{25} \times \frac{35}{42}$

Mixed Review 14

A calculator is not allowed.

1. Find the reciprocal of each number.

a. $1\frac{1}{23}$	**b.** $3\frac{2}{11}$	**c.** 79	**d.** 100	**e.** $\frac{3}{1000}$

2. Divide.

a. $\frac{6}{7} \div \frac{1}{7}$	**b.** $1\frac{9}{20} \div \frac{3}{20}$	**c.** $5 \div \frac{1}{3}$	**d.** $7 \div 1\frac{2}{5}$

3. A sheet of stickers has 48 stickers that each measure 1 1/4 in by 1 1/4 in. Little Hannah starts to stick them on the front cover of a notebook, side by side. The notebook measures 5 1/2 in by 8 1/2 in. How many stickers can she fit on the cover?

4. Fill in the table.

Expression	the terms in it	coefficient(s)	Constants
$2x + 3y$			
$0.9s$			
$2a^4c^5 + 6$			
$\frac{1}{6}f$			

5. The table lists the quantities of some of the ingredients needed to make cakes of various sizes. Fill in the table.

Serves (people)	6	12	18	24	30
butter		1/2 cup			
sugar		1 cup			
eggs		2			
flour		1 1/2 cups			

6. If you make enough cake for 100 people, how much butter, sugar, eggs, and flour are needed? (You can use the table above.)

7. Find the value of the expressions.

a. $900 - \frac{1}{6} \cdot 72$	**b.** $23 + 3^4$	**c.** $\frac{100^3}{100^2}$

8. Marie's age is 4/7 of her brother Tom's age. Tom is 9 years older than Marie. (You can draw a diagram to help.)

a. How old is Marie? Tom?

b. What is the ratio of Marie's age to Tom's age?

9. Write an expression for both the area *and* perimeter of each shape, in simplified form.

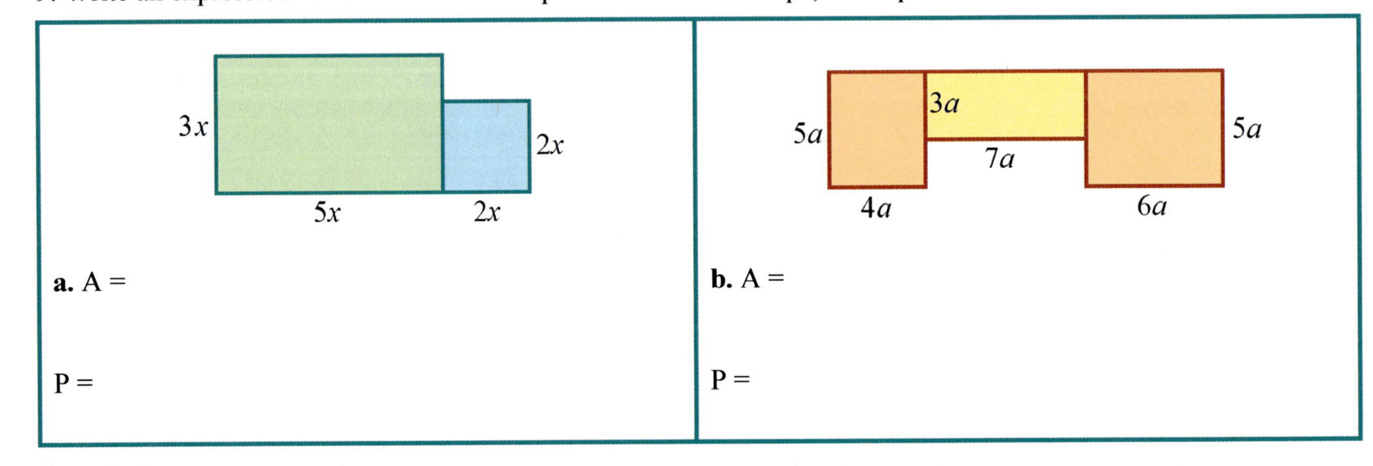

a. A =

P =

b. A =

P =

Geometry Review

1. Explain how the area of the triangle is related to the area of the parallelogram.

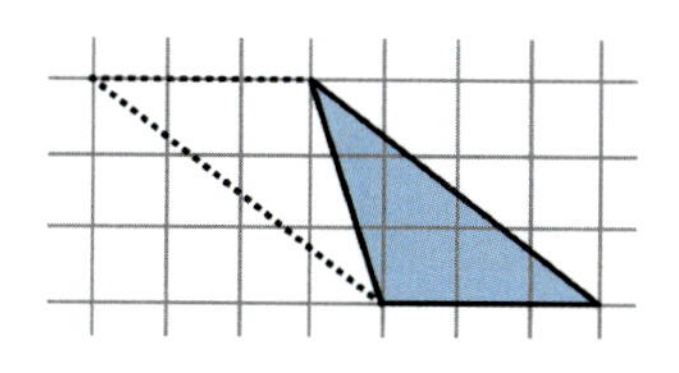

2. Find the area of each quadrilateral in square units.

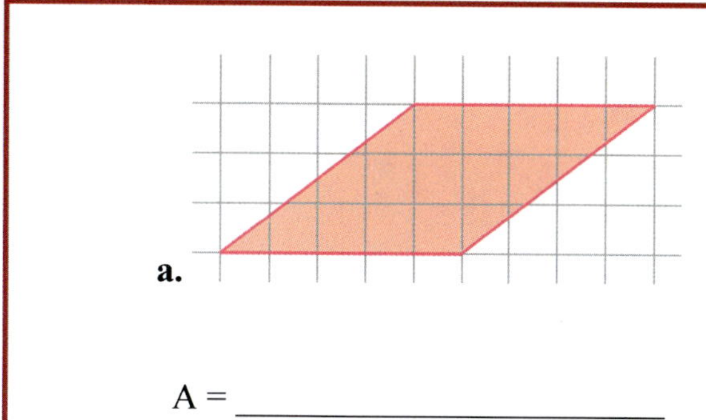

a.

A = ______________________

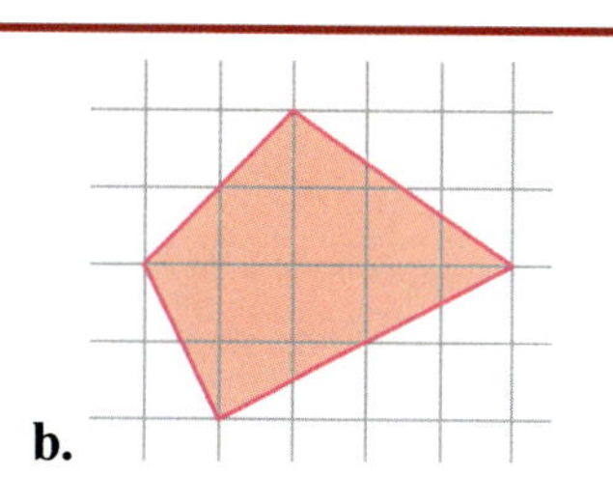

b.

A = ______________________

3. **a.** Jeremy planted a garden in the shape of the diagram at the right. Find the area of Jeremy's garden.

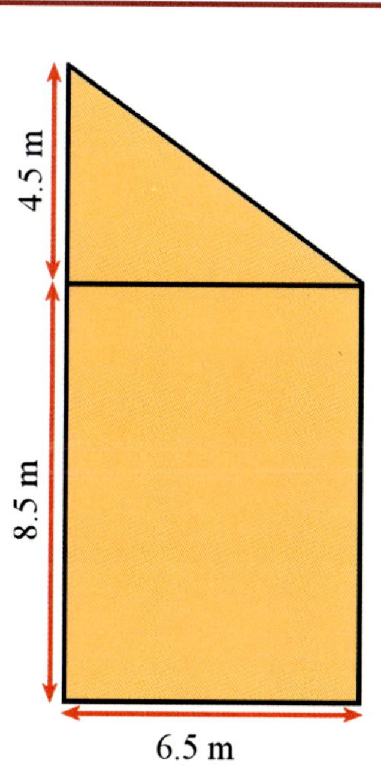

b. Jeremy planted a rectangular section measuring 3.5 m by 3 m with green beans. What percentage of his garden did he plant with green beans?

4. Find the area of this triangle:

a. in square centimeters

b. in square millimeters

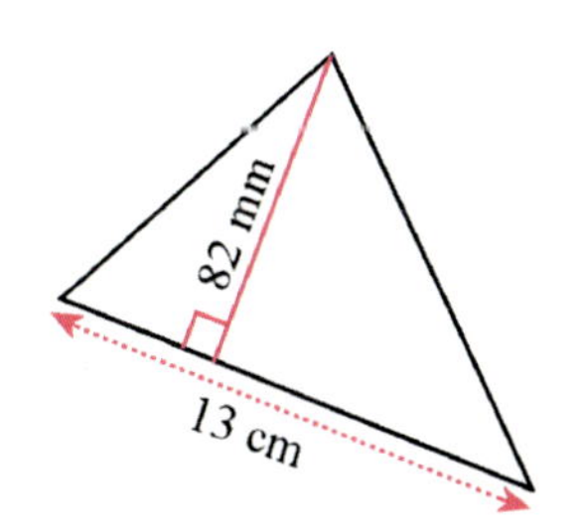

5. Draw a net and calculate the surface area of each solid.

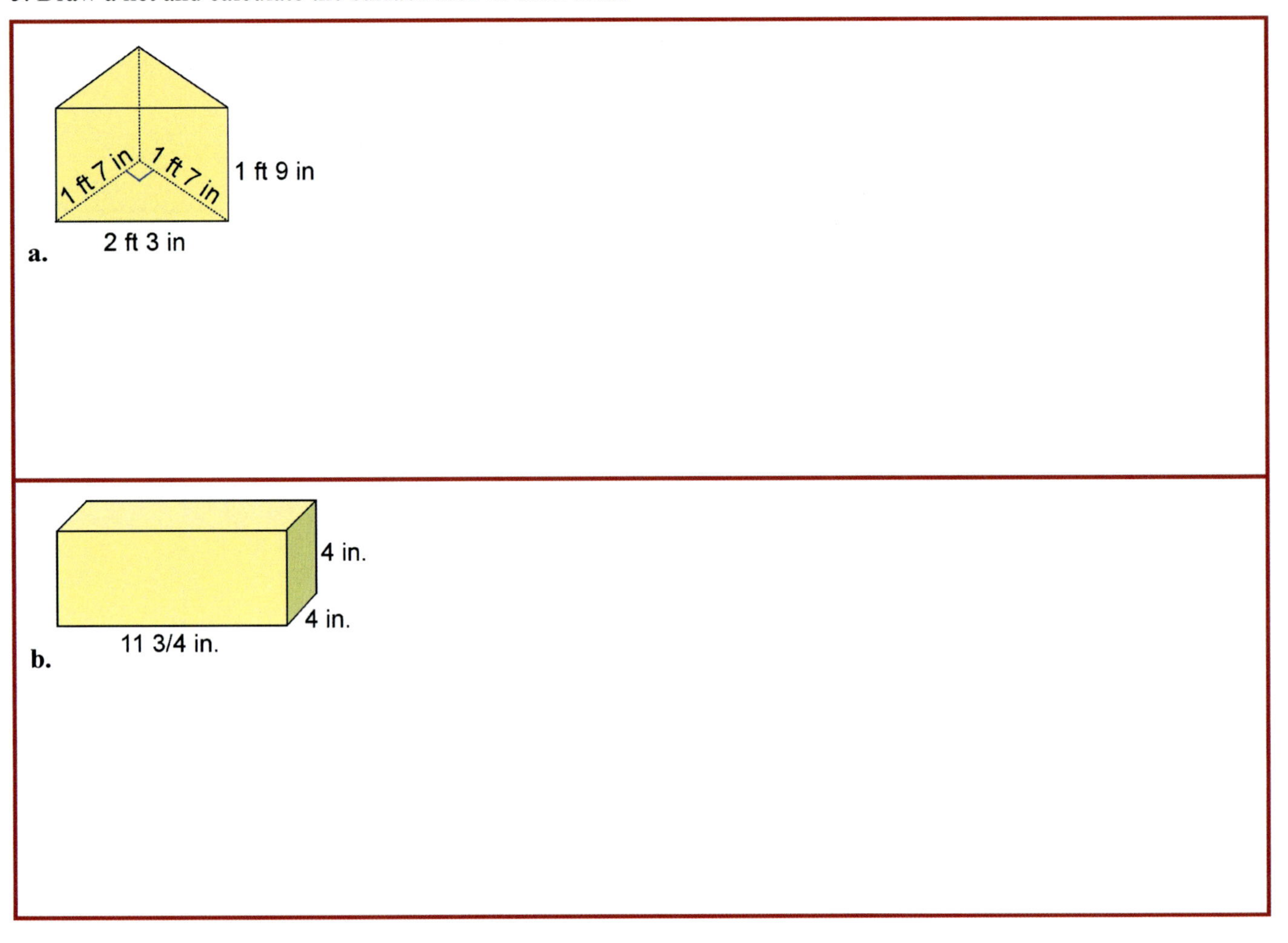

6. What are the names of the solids that can be constructed from these nets?

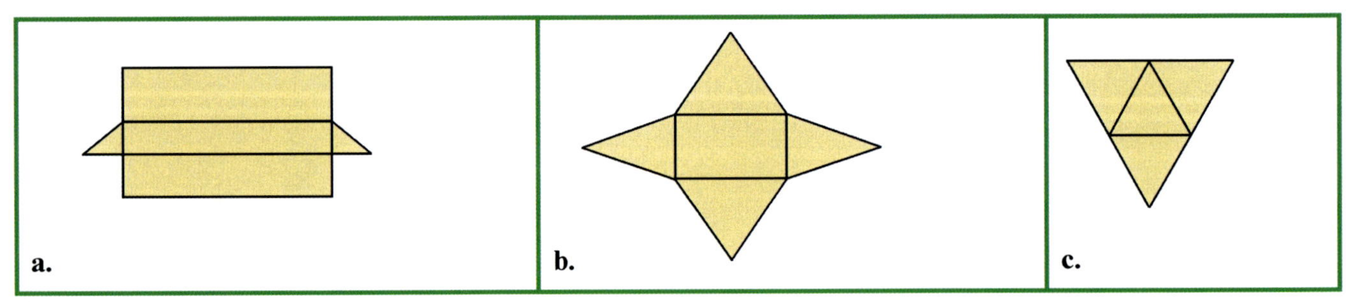

7. What solid can you build from this net?

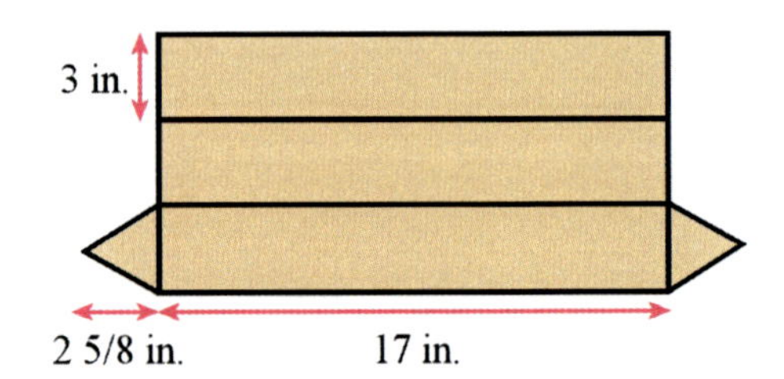

Calculate its surface area.

8. The edges of each little cube measure 1/3 cm. 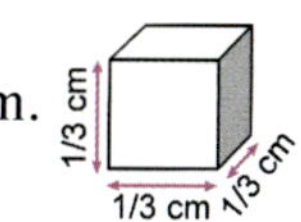

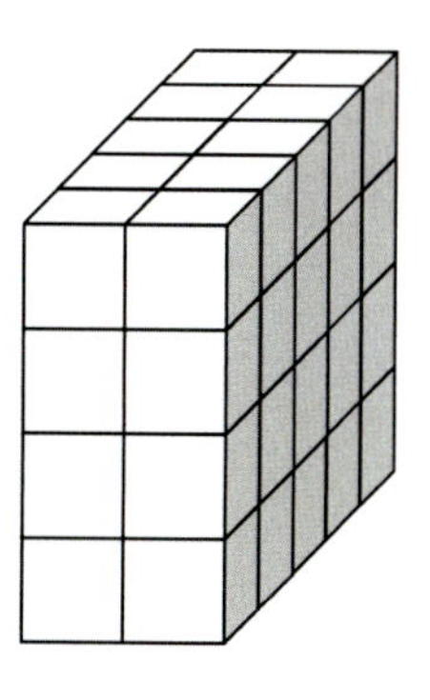

What is the total volume, in cubic meters, of the figure at the right?

9. This building has three stories. Calculate the volume of one story.

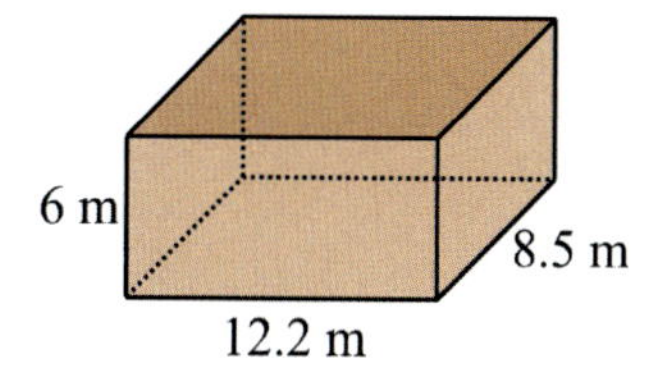

10. An aquarium measures 50 cm × 30 cm on the bottom, and its height is 40 cm.
It is 4/5 filled with water.

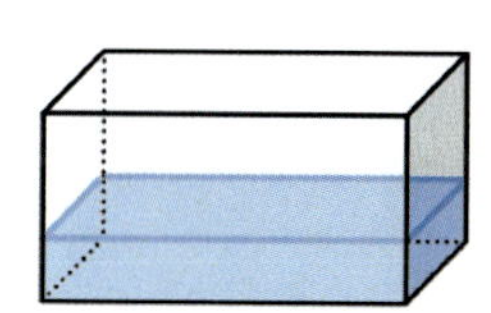

How many cubic centimeters of water is in it?

How many milliliters of water is in it?
(One cubic centimeter is one milliliter.)

How many liters?

Geometry Test

A calculator <u>is not</u> allowed.

1. Measure what you need from the shape, and find its area...
   a. ...in square centimeters, to the nearest square centimeter

   b. ...in square millimeters, to the nearest hundred square millimeters

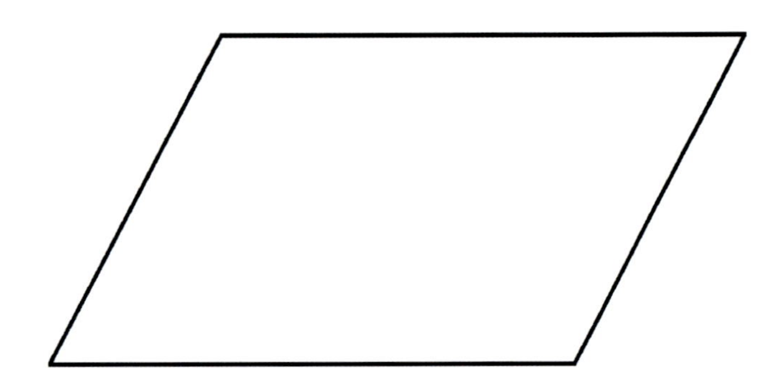

2. Find the area of the quadrilateral in square units.

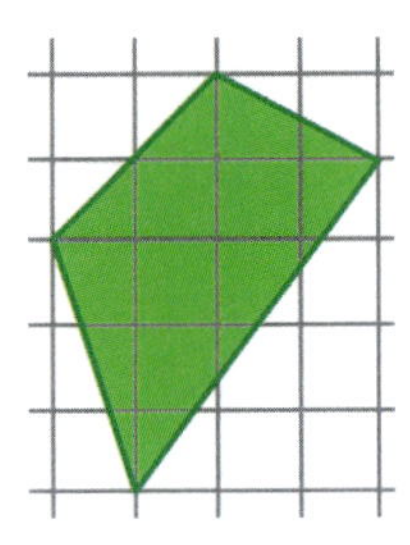

3. **a.** What is this shape called?

   b. Find its area.

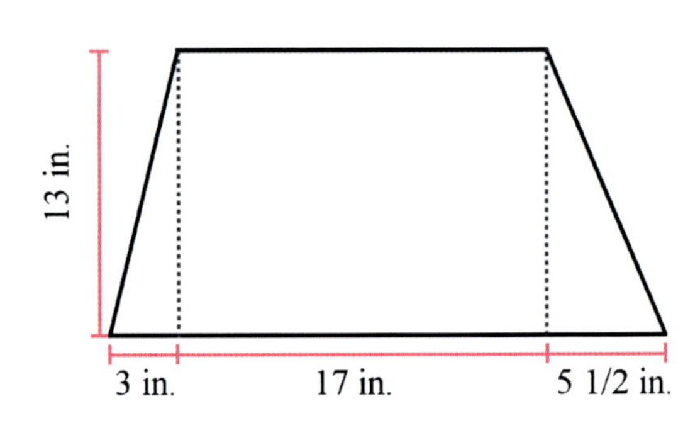

4. The dimensions of this box are 2 ft × 1.5 ft × 1.5 ft. What is the total area of the bottom and side faces of the box (ie. not including the top)?

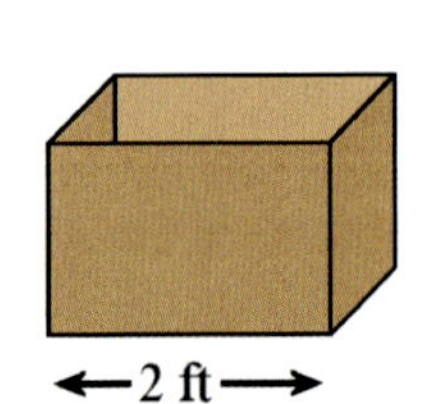

5. The edges of each little cube measure 1/4 in.
   What is the total volume of the figure?

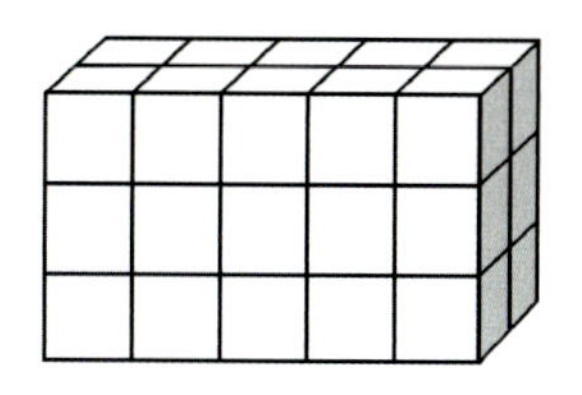

6. A book about how to raise ducks measures 6 1/2 in × 8 in × 3/8 in.
   What is the volume of one book?

7. **a.** What solid can be built from this net?

   b. Calculate its surface area, if each side of the bottom square measures 5 in and the height of each triangle is 4 1/8 in.

8. What is this solid called?

   Sketch its net.

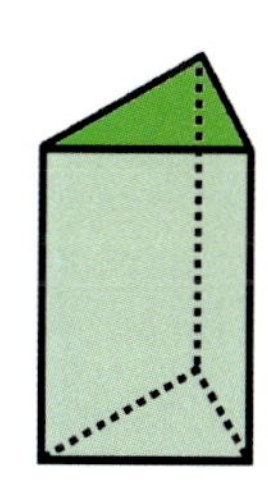

9. The vertices of a triangle are (1, 0), (−2, −4) and (−3, −3).
   Find the area of the triangle.

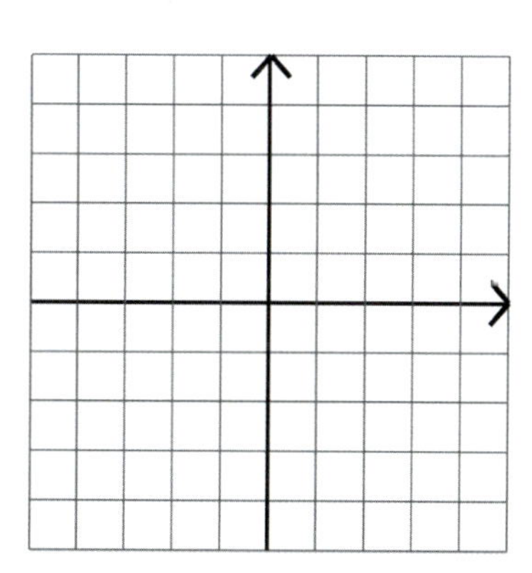

Mixed Review 15

1. A family put 1/3 of 60 pounds of flour into the cellar.
   Then, they gave 3/8 of the remaining flour to a neighbor.
   How much flour did the neighbor get?

2. Multiply.

a. $3 \times 0.3 \times 0.08 =$ ________	**b.** $7 \times 0.2 \times 1.1 =$ ________	**c.** $0.25 \times 10^5 =$ ________
d. $0.0009 \times 8 =$ ________	**e.** $0.002 \times 100 =$ ________	**f.** $3000 \times 0.0007 =$ ________

3. Order the fractions from the smallest to the biggest.

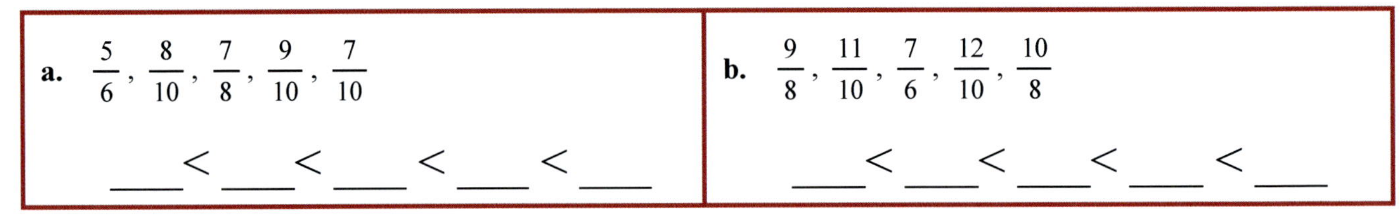

a. $\frac{5}{6}, \frac{8}{10}, \frac{7}{8}, \frac{9}{10}, \frac{7}{10}$	**b.** $\frac{9}{8}, \frac{11}{10}, \frac{7}{6}, \frac{12}{10}, \frac{10}{8}$
____ < ____ < ____ < ____ < ____	____ < ____ < ____ < ____ < ____

4. Convert the measurements into the given units.

   a. 0.9 L = _______ dl = ________ cl = ________ ml

   b. 2,800 m = ________ km = _________ dm = _________ cm

   c. 56 g = _______ dg = ________ cg = ________ mg

5. Convert. Round your answers to 2 decimals in (a) - (d). In (e) and (f) use whole numbers.

a. 76 oz = ____________ lb	**c.** 3.6 gal = ___________ qt	**e.** 2.67 mi = ___________ ft
b. 98 in = ____________ ft	**d.** 0.483 lb = ___________ oz	**f.** 5.09 ft = _____ft _______ in

6. Use ratios to convert the measuring units. 1 kg = 2.2 lb, and 1 in = 2.54 cm.

a. 134 kg into pounds
b. 156 in into centimeters

7. Solve the equations.

a. $0.2m = 6$	**b.** $0.3p = 0.09$	**c.** $y - 1.077 = 0.08$

8. **a.** Draw a picture where there are 2 triangles for each 5 squares, and a total of 21 shapes.

b. The unit rates are:

_______ squares for 1 triangle

_______ triangles for 1 square

9. Add and subtract.

a. $5 + (-8) =$ _______	**b.** $-11 + (-9) =$ _______	**c.** $2 + (-17) =$ _______	**d.** $2 - (-8) =$ _______
$(-5) + 8 =$ _______	$9 - 11 =$ _______	$-3 - 8 =$ _______	$-8 - (-2) =$ _______

10. A figure whose vertices are at $(-5, -3)$, $(-1, -3)$, $(0, -5)$, and $(-7, -5)$ is transformed this way:

1. It is reflected in the x-axis.
2. It is moved four units to the right, five down.
3. It is reflected in the y-axis.

Give the coordinates of its vertices after all three transformations.

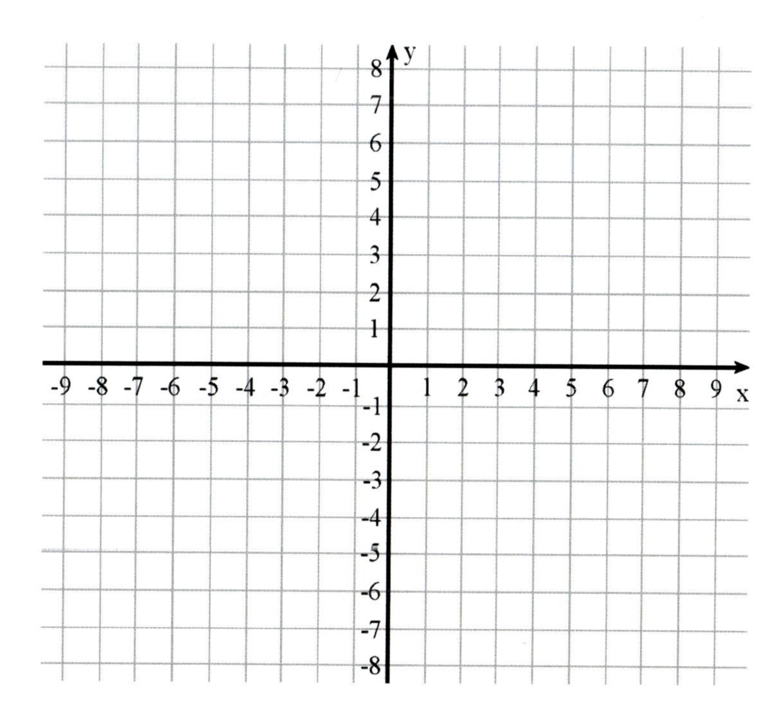

11. Draw a triangle whose vertices are at $(-3, -4)$, $(5, -4)$, and $(2, 7)$.

Draw an altitude to the triangle.

Find its area.

12. A mole is digging a tunnel at the speed of 4 m per hour.

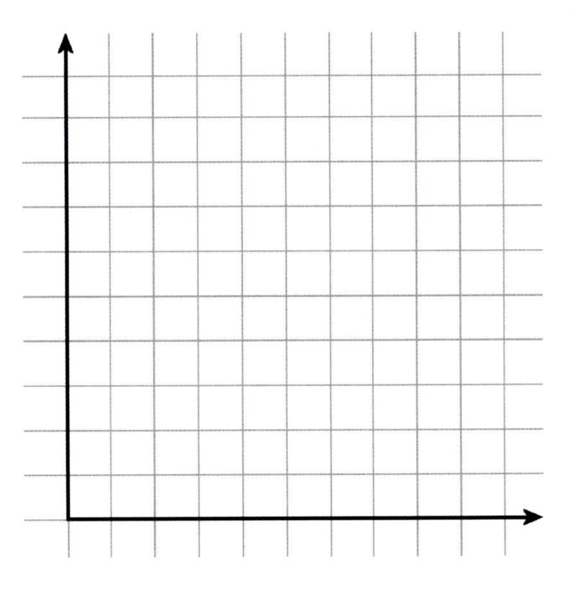

a. Choose a letter variable to represent the time the mole has dug and another to represent the length (distance) of tunnel it has dug.

b. Fill in the table. Plot the points.

time (hours)	0	1	2	3	4	5	6	7	8	9
distance (meters)										

c. Write an equation relating the two variables.

d. Which is the independent variable?

13. Fill in the blank and give an example.

a. Dividing a number by 5 is the same as multiplying it by ____. Example:

b. Dividing a number by $\frac{2}{3}$ is the same as multiplying it by ____. Example:

14. Write as percentages. If necessary, round your answers to the nearest percent.

a. 5/8

b. 6/25

15. Draw a triangle with 55° and 29° angles, and a 6-cm side between those angles.

16. Draw a rhombus with 7.5 cm sides, and one 66° angle.

Puzzle Corner Find the missing factors.

a. $\frac{1}{5} \times \quad = \frac{1}{20}$ **b.** $\frac{1}{5} \times \quad = 2$ **c.** $\frac{5}{6} \times \quad = \frac{1}{3}$

Mixed Review 16

A calculator is not allowed.

1. Move these points four units to the *left*:

   (−5, 1) → (____, ____)

   (−2, −3) → (____, ____)

   (3, −7) → (____, ____)

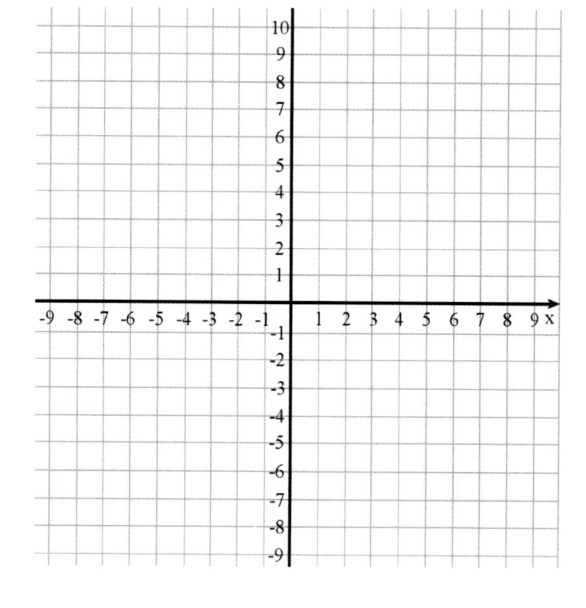

2. Sam and Matt divided a salary of $180 in a ratio of 4:5. Calculate how much each boy got.

3. Find the greatest common factor of the given number pairs.

a. 56 and 70	**b.** 96 and 36

4. Find five numbers that are multiples of both 5 and 9.

5. Solve the equations by thinking logically.

a. 4 × ________ = 0.0012	**b.** 0.2 × ________ = 0.06	**c.** 0.03 × ________ = 30

6. Solve the equations.

a. $0.5x = 30$	**b.** $0.01x = 2$	**c.** $c + 1.1097 = 3.29$

7. The grid represents a board game.
Samantha has game pieces at (−50, 40) and (−50, −25).

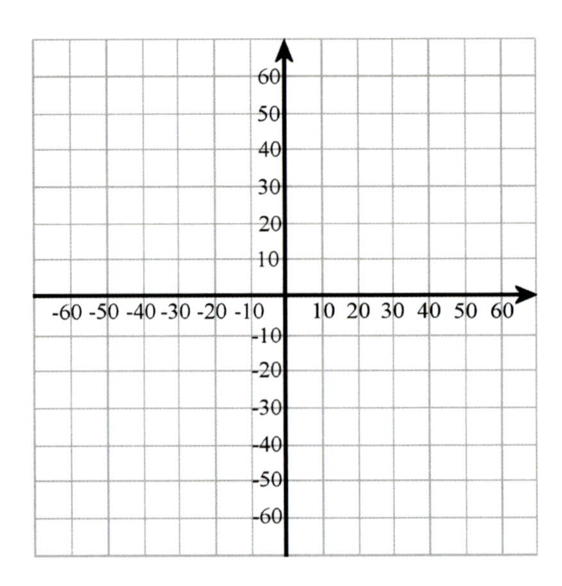

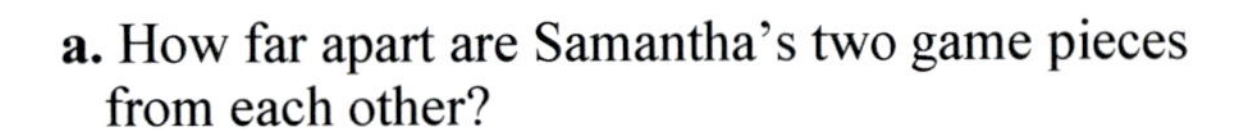

a. How far apart are Samantha's two game pieces from each other?

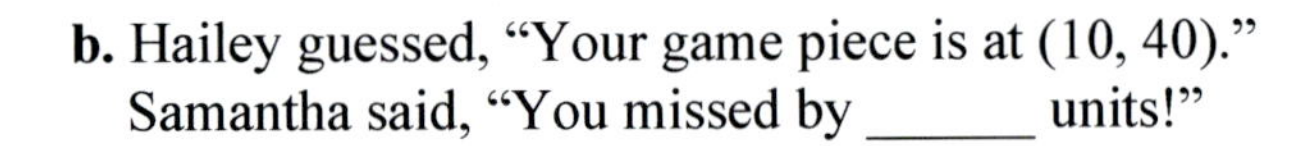

b. Hailey guessed, "Your game piece is at (10, 40)."
Samantha said, "You missed by ______ units!"

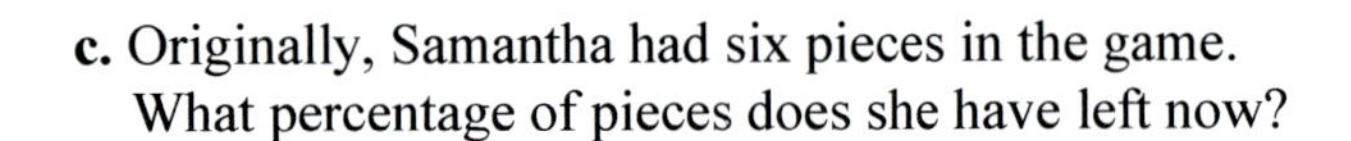

c. Originally, Samantha had six pieces in the game.
What percentage of pieces does she have left now?

8. Find the better deal: an \$18 flash drive is discounted by 15%, and another, \$20 flash drive is discounted by 1/5.

9. Alice had a box of 90 oranges. She gave 3/5 of the oranges to Beatrice.
Then, of what was left, she gave 1/4 to Michael.
How many oranges does Alice have now?
How many oranges did Michael get?

10. Asphalt paving costs \$1,250 for 500 square feet. Fill in the equivalent rates.

$$\frac{\_\_\_\_\_\_}{100\text{ sq. ft.}} = \frac{\_\_\_\_\_\_}{200\text{ sq. ft.}} = \frac{\_\_\_\_\_\_}{500\text{ sq. ft.}} = \frac{\_\_\_\_\_\_}{2{,}000\text{ sq. ft.}} = \frac{\_\_\_\_\_\_}{2{,}400\text{ sq. ft.}}$$

11. Add and subtract.

a. $-2 + (-11) =$ ________	**b.** $-1 + (-7) =$ ________	**c.** $10 - 17 =$ ________	**d.** $7 - (-3) =$ ________
$(-11) + 2 =$ ________	$1 - 7 =$ ________	$-10 - 17 =$ ________	$-3 - (-7) =$ ________

12. Multiply, and shade the grid to illustrate the multiplications.

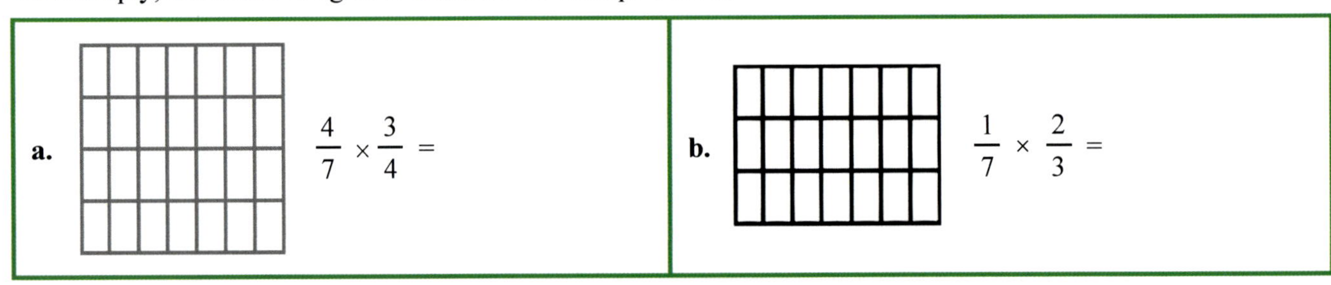

a. $\frac{4}{7} \times \frac{3}{4} =$

b. $\frac{1}{7} \times \frac{2}{3} =$

Statistics Review

1. Are these statistical questions or not? If not, change the question so that it becomes a statistical question.

   a. Which kind of books do the visitors of this library like the best?

   b. How many pages are in the book *How to Solve It* by G. Polya?

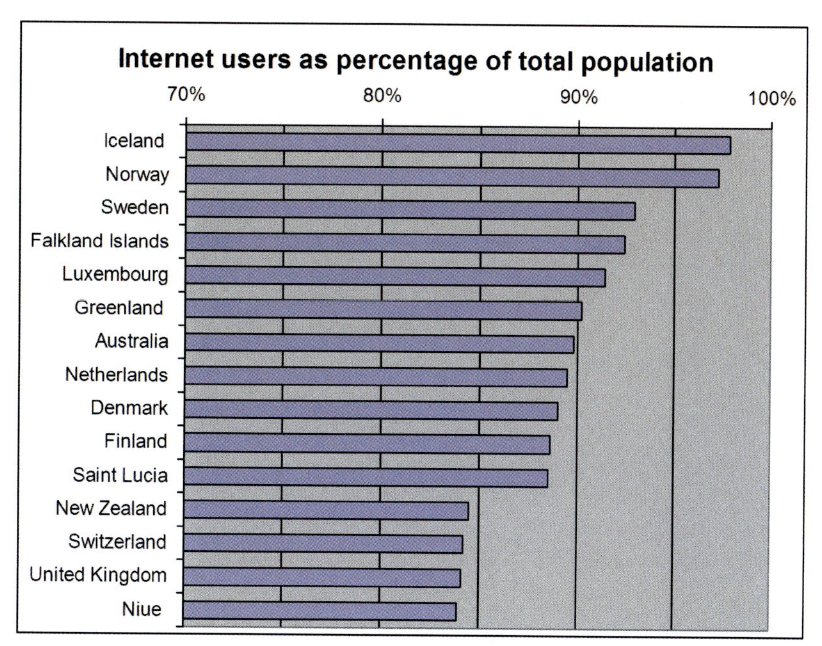

Source: InternetWorldStats.com

2. Fill in, using estimated percentages from the graph. In (c) and (d), round to the nearest tenth of a million.

   a. About ________% of the population of Norway uses the Internet.

   b. About ________% of the population of the United Kingdom uses the Internet.

   c. The population of the Netherlands was approximately 16,847,000 in 2011.

   So there are about _________ million Internet users in the Netherlands.

   d. The population of Finland was about 5,260,000 when these statistics were gathered (2011).

   So there are about _________ million Internet users in Finland.

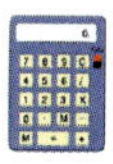

3. **a.** Find the mean, median, and mode.
   Hint: recreate the list of the original data.

   Mean:

   Median:

   Mode:

   b. We notice this distribution has a **gap** at 5. What else can you say about the shape of the distribution?

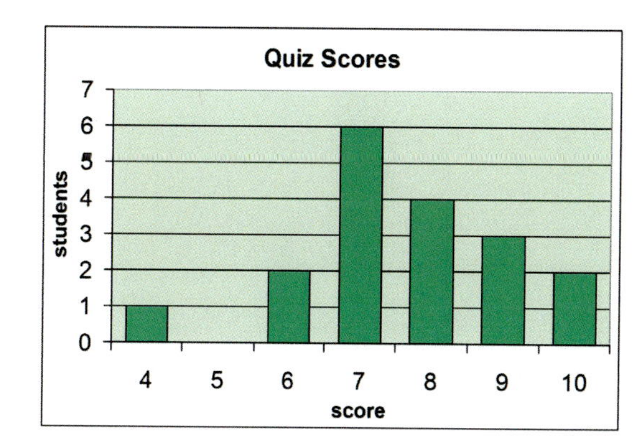

4. **a.** Find the five-number summary and the interquartile range of this data set, and make a boxplot.

2, 5, 5, 6, 6, 7, 7, 7, 8, 8, 8, 9, 12

Minimum: ________

First quartile: ________

Median: ________

Third quartile: ________

Maximum: ________

Interquartile range: ________

b. What could this data be?

5. **a.** Make a stem-and-leaf plot of this data.

78 82 84 75 90 66 77 64 112 84 85

(The height of a group of toddlers, in centimeters.)

b. Find the median.

c. Find the range.

d. The data set has an outlier.
Which number is the outlier?

e. Describe the shape of the distribution.

Stem	Leaf

6. This graph shows the hourly wages in euros per hour of the 89 employees in the Inkypress Print Shop.

a. About what fraction of the people earn 7-8 euros/hour?

b. Describe the shape of the distribution.

c. The mean is 9.66 euros/hour and the median is 8 euros/hour. Which is better in describing the majority's wages in this print shop?

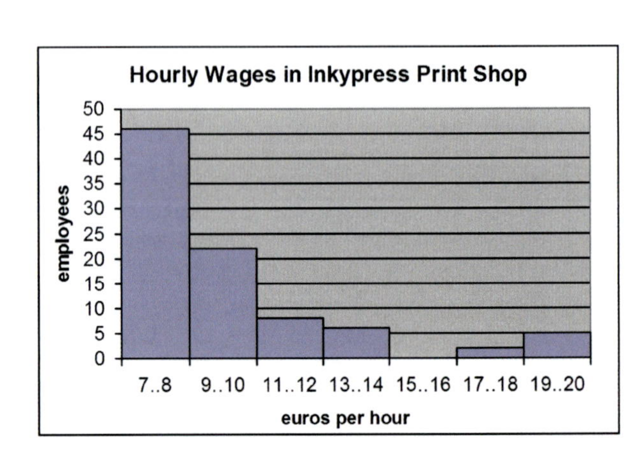

7. **a.** Create a dot plot from this data.

b. Describe the spread of the data.

c. Describe the shape of the distribution.

d. Choose a measure of center to describe the data, and determine its value.

e. Create a histogram. Make four bins.

California Cities Precipitation

City	Average Annual Rainfall (in)
Bishop	5
Bakersfield	6
San Diego	11
Fresno	11
Long Beach	13

City	Average Annual Rainfall (in)
Los Angeles	13
San Francisco	13
Stockton	14
Santa Maria	14
Santa Barbara	17

City	Average Annual Rainfall (in)
Sacramento	18
Redding	34
Eureka	38
Mount Shasta	39
Blue Canyon	68

Dot plot:

Histogram:

Rainfall (in)	Frequency

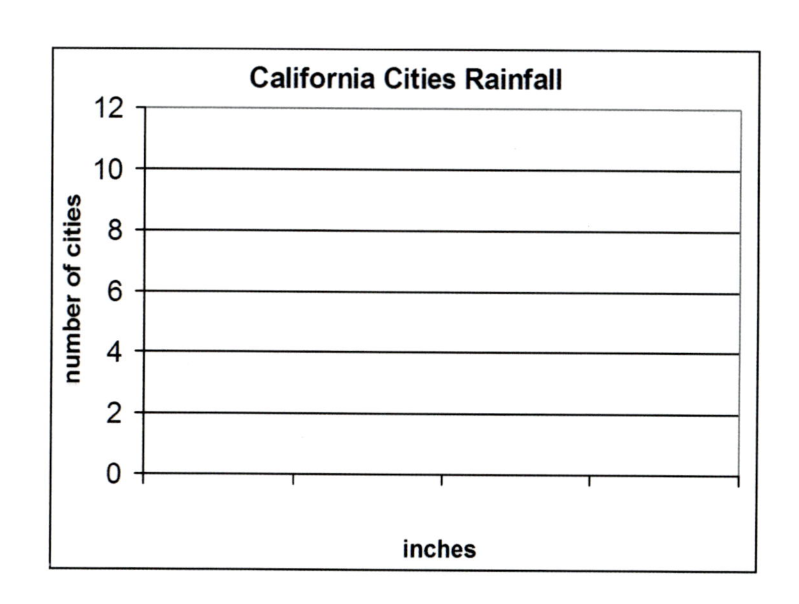

Statistics Test

A calculator <u>is</u> allowed.

1. Calculate the mean, median, mode, and range—if possible—for these data sets.

   a. 12, 15, 11, 18, 20, 15, 16

   mean _________ median _________ mode _________ range _________

   b. duck, cow, horse, horse, horse, cat, cat, dog, dog, dog

   mean _________ median _________ mode _________ range _________

2. The following are the points for two math quizzes for a 7th grade class.

   a. Make bar graphs from the data.

   b. Describe the shape of each distribution.

   c. Choose a measure of center to describe the distributions, and determine its value for both quizzes.

   d. Which quiz went better overall?

Quiz 1	
Points	**Students**
5	7
6	8
7	6
8	3
9	0
10	0

Quiz 2	
Points	**Students**
5	1
6	2
7	5
8	8
9	5
10	3

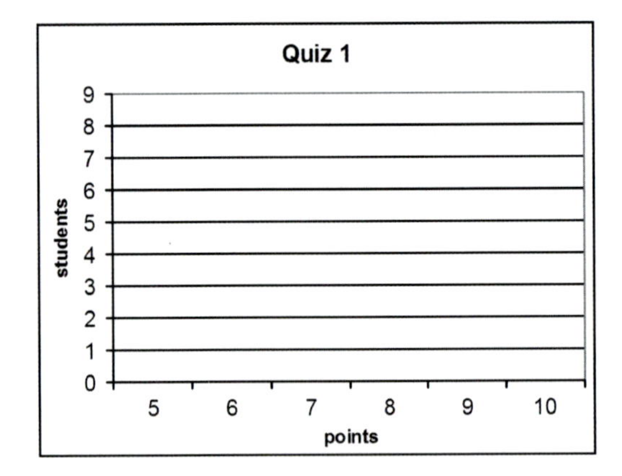

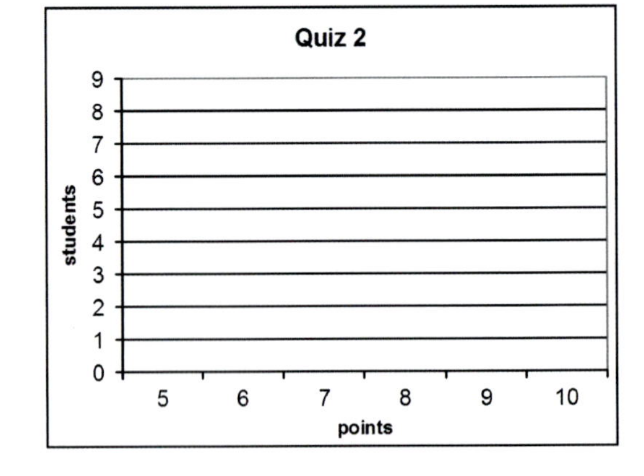

Quiz 1:

Shape of the distribution: ____________________

____

Measure of center: ________________________

Quiz 2:

Shape of the distribution: ____________________

____

Measure of center: ________________________

3. **a.** Make a stem-and-leaf plot of this data.

114 128 132 127 122 127 130 119 120 121 125
(Results of a high jump contest boys, in centimeters)

Stem	Leaf

b. Find the median.

c. What is the interquartile range?

4. Make a boxplot from this data:

89 92 95 96 99 103 105 106 106 109 109 110 112 114 117 118 124

(birth weight in grams of Momma Cat's three litters of kittens)

Mixed Review 17

1. Find the greatest common factor of the given number pairs.

a. 87 and 36

b. 96 and 16

2. Find the least common multiple of the given number pairs.

a. 6 and 12

b. 8 and 12

3. First, find the GCF of the numbers. Then factor the expressions using the GCF.

a. The GCF of 72 and 12 is ______ 12 + 72 = ____ (____ + ____)	**b.** The GCF of 42 and 66 is ______ 42 + 66 = ____ (____ + ____)

4. **a.** The points (−8, 7), (−5, 3), and (4, 0) are vertices of a triangle. Draw the triangle.

b. Move the triangle five units down *and* three units to the right. Notice there are *two* movements! Write the coordinates of the moved vertices.

(−8, 7) → (_____ , _____)

(−5, 3) → (_____ , _____)

(4, 0) → (_____ , _____)

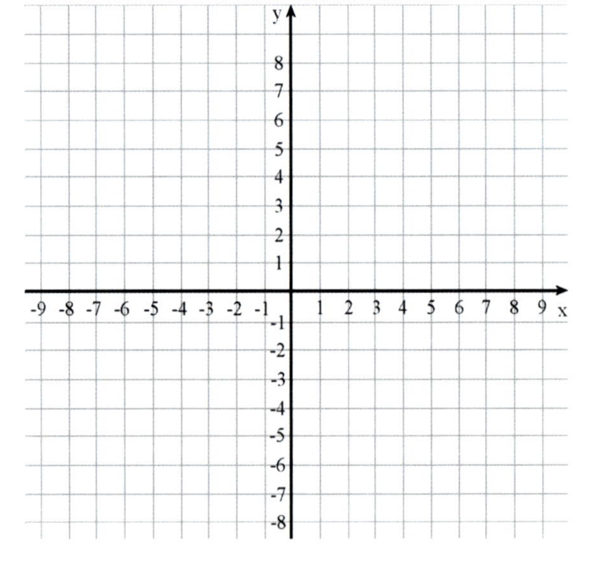

5. Write an equation for each situation—even if you could easily solve the problem without an equation! Lastly, *solve* the equation you wrote.

a. The area of a rectangle is 304 m^2 and one of its sides is 19 m. How long is the other side?
b. Mike weighed five identical books on the scales. They weighed 6.7 kg. What was the weight of one book?

6. Simplify the expressions.

a. $z \cdot z \cdot z \cdot 7$	**b.** $8 \cdot a \cdot 3 \cdot b \cdot 10$
c. $2 + x + x + x + x$	**d.** $5t - 2t + 6$

7. Add or subtract the fractions. Give your answer as a mixed number.

a. $\frac{5}{11} + \frac{1}{2} + \frac{5}{6}$

b. $3\frac{11}{12} - \frac{5}{10} + \frac{1}{4}$

8. What part of a whole pizza is two-thirds of nine-tenths of a pizza?

9. A piglet is born weighing 3 lb 4 oz. If it gains approximately 7 1/3 ounces per day during its 12-day nursing period, then how much will it weigh at weaning (the end of the nursing period)?

10. A string that is 5 3/4 inches long is cut into four equal pieces. How long are the pieces?

11. Simplify. In (e), write using a number.

a. $|9|$ **b.** $|-3|$ **c.** $|0|$ **d.** $-(-28)$ **e.** the opposite of -7

12. Write an addition or subtraction sentence.

a. You are at $^{-}12$. You jump 7 steps to the right. You end up at ______.

b. You are at 2. You jump 8 steps to the left. You end up at ______.

13. On a separate sheet of paper, draw a right triangle with an *area* of 8 square inches.

14. Find the total area of the boat and its sail.

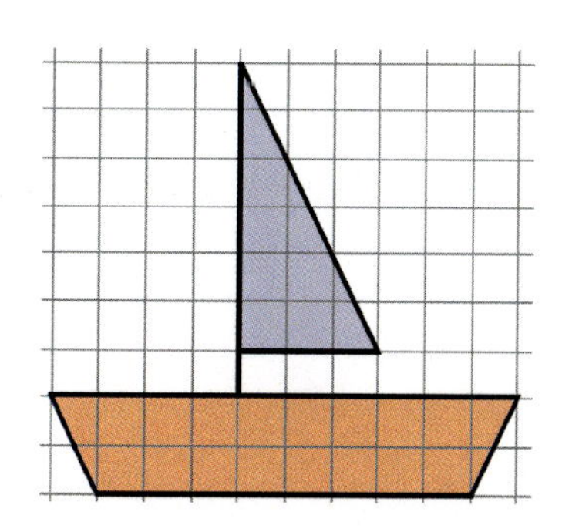

15. Find the area of the yellow shaded figure at the right.

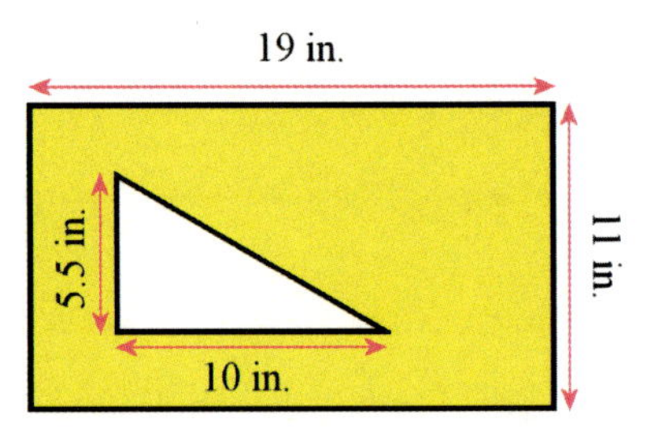

16. **a.** Name the three warmest and the three coldest months in Boston.

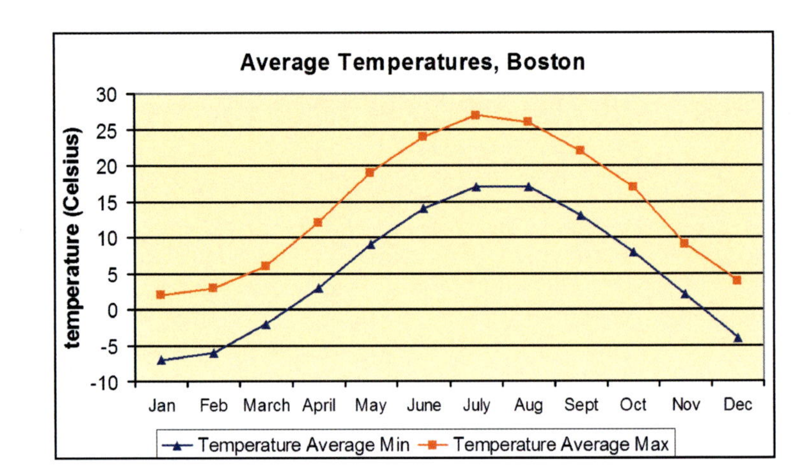

b. Now look at the maximum temperatures. What is the temperature difference between the coldest and the warmest month?

c. About how much is the difference in maximum and minimum temperatures in August?

In January?

Puzzle Corner Be a teacher-detective: how did the children come up with these answers?

a. Jerry cannot figure out what went wrong:

$$\frac{2}{7} \div 1\frac{3}{4} = 6\frac{1}{8}$$

$$2\frac{1}{3} \div \frac{2}{5} = \frac{6}{35}$$

$$1\frac{1}{5} \div 2\frac{2}{3} = 2\frac{2}{9}$$

What error did Jerry make each time?

b. Emily has something fishy going on here:

$$\frac{4}{5} \div 1\frac{1}{2} = 1\frac{3}{5}$$

$$2\frac{1}{3} \div \frac{1}{4} = 1\frac{1}{3}$$

$$1\frac{1}{5} \div 2\frac{2}{3} = \frac{3}{10}$$

What error did Emily make each time?

Mixed Review 18

A calculator is not allowed.

1. Find the perimeter and the area of this triangle.

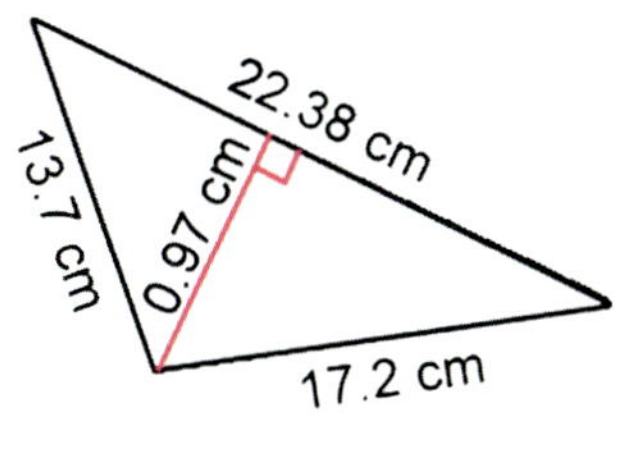

2. The distance from Ben's home to his workplace is only 0.7 miles.

   a. What is it in feet?

   b. Ben walks to work 4/5 of his workdays, and the rest of the time he rides a bike. Calculate how many miles he ends up walking in a year going to work. Assume that he works 48 weeks in a year, 5 days a week.

3. Convert the measurements into the given units.

	m	**dm**	**cm**	**mm**
a. 7.82 m	7.82			
b. 109 mm				109 mm

4. Divide, and give your answer as a decimal. If necessary, round the answers to three decimal digits.

a. $17.54 ÷ 3	**b.** 2.4 ÷ 0.05

5. **a.** What is the volume of a shoe box measuring 25 cm by 18 cm by 12 cm?

b. Sketch the net of the shoe box.

c. Calculate the surface area of the shoe box.

6. Write an expression.

a. 5 less than x to the 5th power.

b. The quantity 2 minus x, cubed.

c. 2 times the sum of 10 and y.

d. The difference of s and 2, divided by s squared.

7. Find the value of the expression in question 6a above, if x has the value 2.

8. Factor these sums (write them as products).

a. $56x + 14 =$ ____ (____ + ____)	**b.** $18u + 60 =$ ____ (____ + ____)

9. Solve the equations.

a. $y \div 50 = 60 \cdot 2$	**b.** $3x - x = 3 + 7$	**c.** $7x = 50$
=	=	=
=	=	=
=	=	=

10. Solve the inequality $x - 12 > 6$ in the set $\{11, 13, 15, 17, 19, 21, 23\}$.

11. Mark the following numbers on this number line that starts at 0 and ends at 2.

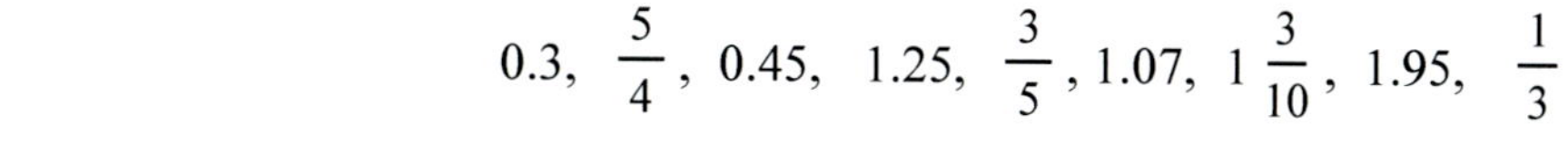

$0.3, \ \frac{5}{4}, \ 0.45, \ 1.25, \ \frac{3}{5}, \ 1.07, \ 1\frac{3}{10}, \ 1.95, \ \frac{1}{3}$

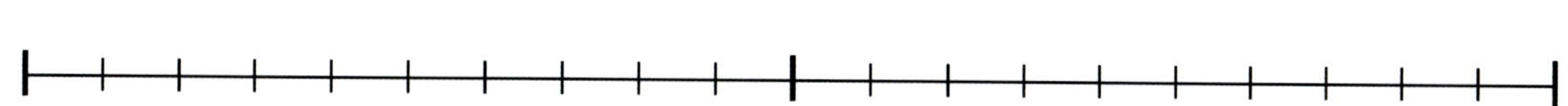

12. Write these fractions as decimals. Give your answers to three decimal digits.

a. $\frac{5}{4} =$	**b.** $\frac{6}{7} =$	**c.** $\frac{19}{16} =$

13. A puzzle measures 8 1/2 inches by 10 inches. Calculate the area of the puzzle in square centimeters, using the fact that 1 inch = 2.54 cm.

14. Oats cost $0.92 per pound. Eric bought 2 3/4 lb. Calculate the total cost of Eric's purchase.

15. Make a stem-and-leaf plot from the following data:

heart rates of a group of 13-year olds after doing jumping jacks for 30 seconds:

159 162 145 175 155 163 160 140 158 190 172 162 152 163 148 150

Grade 6 End-of-the-Year Test

This test is quite long, because it contains lots of questions on all of the major topics covered in the *Math Mammoth Grade 6 Complete Curriculum.* Its main purpose is to be a diagnostic test—to find out what the student knows and does not know. The questions are quite basic and do not involve especially difficult word problems.

Since the test is so long, I do not recommend that you have the student do it in one sitting. You can break it into 3-5 parts and administer them on consecutive days, or perhaps on morning/evening/morning/evening. Use your judgment.

A calculator is not allowed, except on the page about measuring units.

The test is evaluating the student's ability in the following content areas:

- exponents, expanded form, and rounding
- writing and simplifying expressions
- the distributive property
- the concept of an equation and solving simple equations
- the concept of inequality
- all operations with decimals
- conversions between measuring units
- basic ratio concepts
- the concept of percentage, finding percentages, finding the percent of number
- prime factorization, the greatest common factor, and the least common multiple
- division of fractions
- basic concepts related to integers
- addition and subtraction of integers
- the area of triangles, parallelograms, and polygons
- surface area and nets
- the volume of rectangular prisms
- describing statistical distributions
- measures of center
- statistical graphs

Instructions to the teacher:
In order to continue with the *Math Mammoth Grade 7 Complete Worktext,* I recommend that the student score a minimum of 80% on this test, and that the teacher or parent review with the student any content areas in which the student may be weak. Students scoring between 70% and 80% may also continue with grade 7, depending on the types of errors (careless errors or not remembering something, versus a lack of understanding). Use your judgment.

My suggestion for points per item is as follows. The total is 194 points. A score of 155 points is 80%.

Question #	Max. points	Student score
Basic Operations		
1	2 points	
2	3 points	
3	2 points	
4	2 points	
	subtotal	/ 9
Expressions and Equations		
5	4 points	
6	2 points	
7	2 points	
8	1 point	
9	2 points	
10	2 points	
11	2 points	
12	2 points	
13	2 points	
14	2 points	
15	1 point	
16	2 points	
17	2 points	
18	2 points	
19	4 points	
	subtotal	/ 32
Decimals		
20	2 points	
21	2 points	
22	1 point	
23	2 points	
24	2 points	
25	1 point	

Question #	Max. points	Student score
Decimals, cont.		
26	2 points	
27	2 points	
28a	1 point	
28b	2 points	
29	3 points	
	subtotal	/ 20
Measuring Units		
30	3 points	
31	1 point	
32	2 points	
33	3 points	
34	6 points	
35	4 points	
	subtotal	/ 19
Ratio		
36	2 points	
37	2 points	
38	2 points	
39	2 points	
40	2 points	
41	2 points	
42	2 points	
	subtotal	/ 14
Percent		
43	3 points	
44	4 points	
45	2 points	
46	2 points	
47	2 points	
	subtotal	/13

Question #	Max. points	Student score
Prime Factorization, GCF, and LCM		
48	3 points	
49	2 points	
50	2 points	
51	2 points	
52	2 points	
	subtotal	/11
Fractions		
53	3 points	
54	2 points	
55	2 points	
56	2 points	
57	3 points	
58	3 points	
	subtotal	/15
Integers		
59	2 points	
60	2 points	
61	2 points	
62	4 points	
63	5 points	
64	6 points	
65	4 points	
	subtotal	/25

Question #	Max. points	Student score
Geometry		
66	1 point	
67	1 point	
68	3 points	
69	4 points	
70	2 points	
71a	1 point	
71b	3 points	
72	4 points	
73a	2 points	
73b	2 points	
	subtotal	/23
Statistics		
74a	2 points	
74b	1 point	
74c	2 points	
75a	1 point	
75b	1 point	
76a	2 points	
76b	1 point	
76c	1 point	
76d	2 points	
	subtotal	/13
	TOTAL	/194

Math Mammoth End-of-the-Year Test - Grade 6

Basic Operations

1. Two kilograms of ground cinnamon is packaged into bags containing 38 g each. There will also be some cinnamon left over. How many bags will there be?

2. Write the expressions using an exponent. Then solve.

a. $2 \times 2 \times 2 \times 2 \times 2$

b. five cubed

c. ten to the seventh power

3. Write in normal form (as a number).

a. $7 \times 10^7 + 2 \times 10^5 + 9 \times 10^0$

b. $3 \times 10^8 + 4 \times 10^6 + 5 \times 10^5 + 1 \times 10^2$

4. Round to the place of the underlined digit.

a. $6,29\underline{9},504 \approx$ ____________________

b. $6,609,\underline{9}42 \approx$ ____________________

Expressions and Equations

5. Write an expression.

a. 2 less than s.

b. The quantity $7 + x$, squared.

c. Five times the quantity $y - 2$.

d. The quotient of 4 and x^2.

6. Evaluate the expressions when the value of the variable is given.

a. $40 - 8x$ when $x = 2$	**b.** $\frac{65}{p} \cdot 3$ when $p = 5$

7. Write an expression for each situation.

a. You bought m yogurt cups at \$2 each and paid with \$50. What is your change?

b. The area of a square with the side length s.

8. Write an expression for the total length of the line segments, and simplify it.

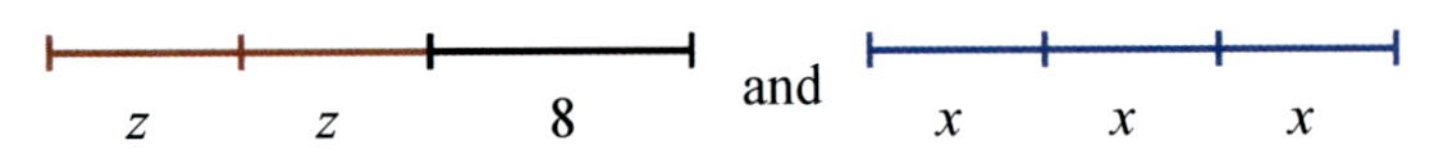

9. Write an expression for the perimeter of the figure, and simplify it.

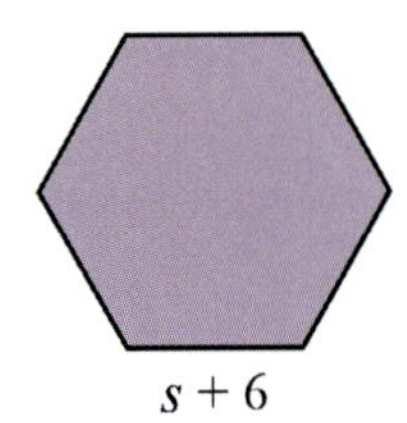

10. Write an expression for the area of the figure, and simplify it.

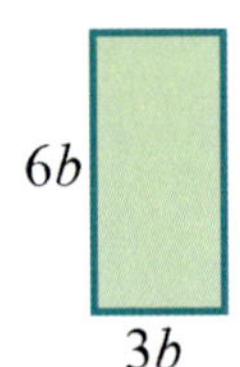

11. Simplify the expressions.

a. $9x - 6x$	**b.** $w \cdot w \cdot 7 \cdot w \cdot 2$

12. Multiply using the distributive property.

a. $7(x + 5) =$	**b.** $2(6p + 5) =$

13. Find the missing number in the equations.

a. ____ $(6x + 5) = 12x + 10$	**b.** $5(2h +$ ____ $) = 10h + 30$

14. Solve the equations.

a. $\frac{x}{31} = 6$	**b.** $a - 8.1 = 2.8$

15. Which of the numbers 0, 1, 2, 3 or 4 make the equation $\frac{8}{y^2} = 2$ true?

16. Write an equation even if you could easily solve the problem without an equation. Then solve the equation.

The value of a certain number of quarters is 1675 cents. How many quarters are there?

17. Write an inequality for each phrase. You will need to choose a variable to represent the quantity in question.

a. Eat at most 5 pieces of bread.

b. You have to be at least 21 years of age.

18. Write an inequality that corresponds to the number line plot.

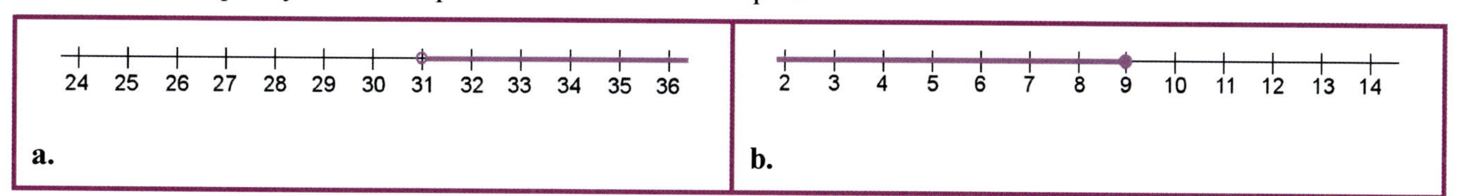

a.

b.

19. A car is traveling at a constant speed of 80 kilometers per hour. Consider the variables of time (t), measured in hours, and the distance traveled (d), measured in kilometers.

a. Fill in the table.

t (hours)	0	1	2	3	4	5	6
d (km)							

b. Plot the points on the coordinate grid.

c. Write an equation that relates t and d.

d. Which of the two variables is the independent variable?

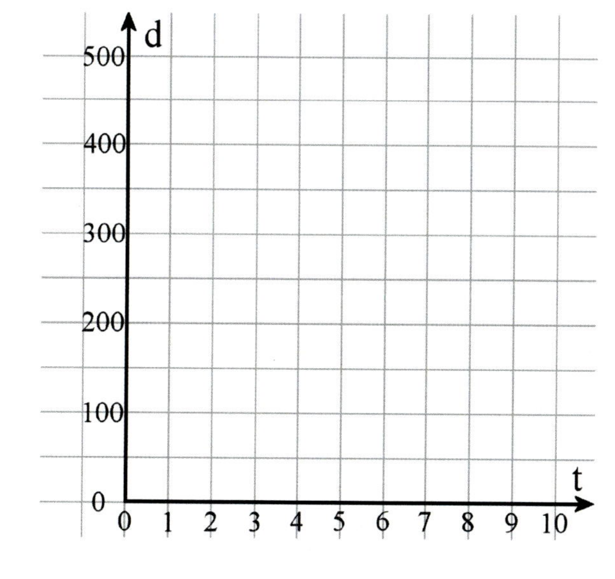

Decimals

20. Write as decimals.

a. 13 millionths

b. 2 and 928 ten-thousandths

21. Write as fractions or mixed numbers.

a. 0.00078

b. 2.000302

22. Find the value of the expression $x + 0.07$ when x has the value 0.0002.

23. Calculate mentally.

a. $0.8 \div 0.1 =$	**b.** $0.06 \times 0.008 =$

24. **a.** Estimate the answer to 7.1×0.0058.

b. Calculate the exact answer.

25. What number is 22 ten-thousandths more than 1 1/2?

26. Multiply or divide.

a. $10^5 \times 0.905 =$	**b.** $24 \div 10^4 =$

27. Divide, and give your answer as a decimal. If necessary, round the answers to three decimal digits.

a. $175 \div 0.3$	**b.** $\frac{2}{9}$

28. Annie bought 3/4 kg of cocoa powder, which cost $12.92 per kg.

a. Estimate the cost.

b. Find the exact amount she had to pay.

29. Alyssa and Anna bought three toy cars for their three cousins from an online store. The price of each car was $3.85. A shipping fee of $4.56 was added to the total cost. The two girls shared the total cost equally. How much did each girl pay?

Measuring Units *A calculator is allowed in this section.* 

1 mile = 5,280 feet 1 mile = 1,760 yards	1 ton = 2,000 lb 1 lb = 16 oz	1 gal = 4 qt 1 qt = 2 pt 1 pt = 16 fl oz

30. Convert to the given unit. Round your answers to two decimals, if needed. 

a. 178 fl oz = __________ qt	**b.** 0.412 mi = __________ ft	**c.** 1.267 lb = ___________ oz

31. How many miles is 60,000 inches? 

32. A big coffee pot makes 2 quarts of coffee.
How many 6-ounce servings can you get from that?

33. A pack of 36 milk chocolate candy bars costs $23.20. Each bar weighs 1.55 oz.
Calculate how much one pound of these chocolate bars would cost (price per pound). 

34. Convert the measurements. You can write the numbers in the place value charts to help you.

a. 39 dl = __________ L **b.** 15,400 mm = __________ m

c. 7.5 hm = __________ cm **d.** 597 hl = __________ L

e. 7.5 hg = __________ kg **f.** 32 g = ___________ cg

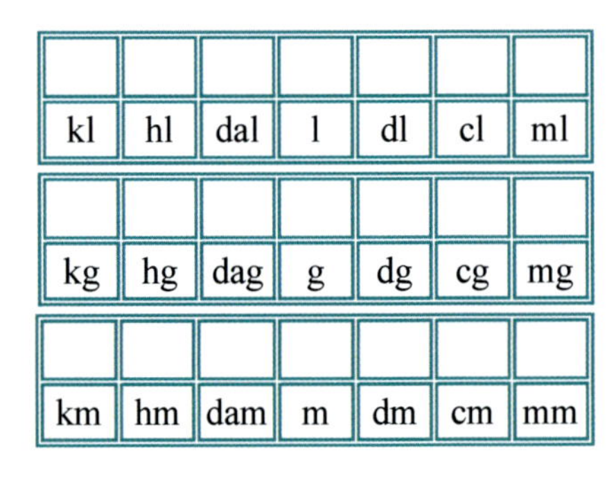

kl	hl	dal	l	dl	cl	ml

kg	hg	dag	g	dg	cg	mg

km	hm	dam	m	dm	cm	mm

35. **a.** One brick is 215 mm long. How many of these bricks, put end to end, will cover a 5.15 meter wall? 

b. Calculate the answer to the previous question again, assuming 1 cm of mortar is laid between the bricks.

Ratio

36. **a.** Draw a picture where there are a total of ten squares, *and* for each two squares, there are three triangles.

b. Write the ratio of squares to all triangles, and simplify this ratio to the lowest terms.

37. Write ratios of the given quantities. Then, simplify the ratios. You will need to *convert* one quantity so it has the same measuring unit as the other.

a. 3 kg and 800 g	**b.** 2.4 m and 100 cm

38. Express these rates in lowest terms.

a. $56 : 16 kg	**b.** There are six teachers for every 108 students.

39. Change to unit rates.

a. $20 for five T-shirts	**b.** 45 miles in half an hour

40. **a.** It took 7 hours to mow four equal-size lawns. At that rate, how many lawns could be mowed in 35 hours? You can use the table below to help.

Lawns					
Hours					

b. What is the unit rate?

41. Joe and Mick worked on a project unequally. They decided to divide their pay in a ratio of 3:4 (3 parts for Joe, 4 parts for Mick). The total pay was $180. Calculate how much Mick got.

42. Use the given ratios to convert the measuring units. If necessary, round the answers to three decimal digits.

a. Use $1 = \dfrac{1.6093\text{ km}}{1\text{ mi}}$ and convert 7.08 miles to kilometers.

7.08 mi =

b. Use $1 = \dfrac{1\text{ qt}}{0.946\text{ L}}$ and convert 4 liters to quarts.

4 L =

Percent

43. Write as percentages, fractions, and decimals.

a. ______% $= \dfrac{35}{100} =$ ______	**b.** 9% $= \dfrac{\quad}{\quad} =$ ______	**c.** ______% $= \dfrac{\quad}{\quad} = 1.05$

44. Fill in the table, using mental math.

	510
1% of the number	
5% of the number	
10% of the number	
30% of the number	

45. A pair of roller skates is discounted by 40%. The normal price is $65.
What is the discounted price?

46. A store has sold 90 notebooks, which is 20% of all the notebooks they had.
How many notebooks did the store have at first?

47. Janet has read 17 of the 20 books she borrowed from the library.
What percentage of the books she borrowed has she read?

Prime Factorization, GCF, and LCM

48. Find the prime factorization of the following numbers.

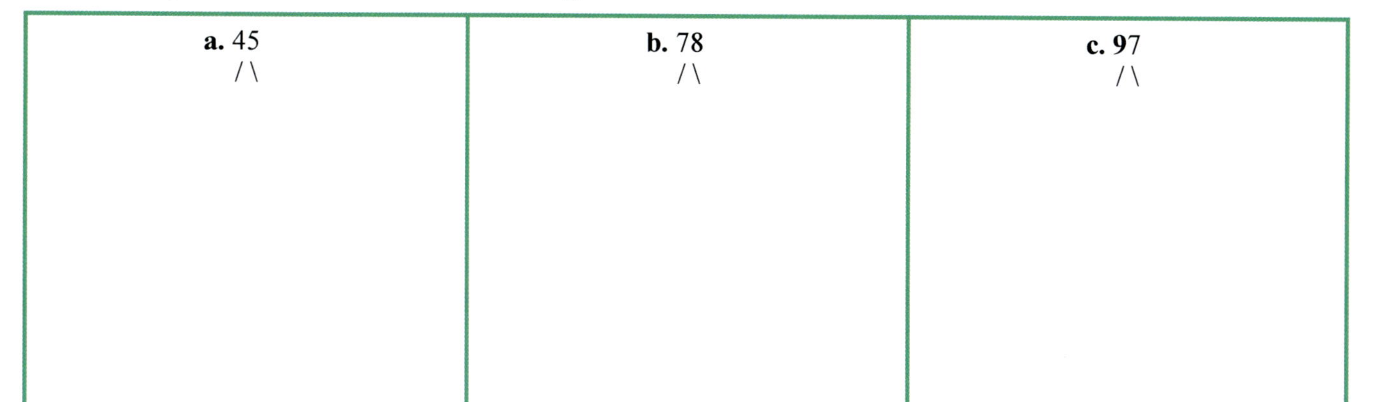

49. Find the least common multiple of these number pairs.

a. 2 and 8	**b.** 9 and 6

50. Find the greatest common factor of these number pairs.

a. 30 and 16	**b.** 45 and 15

51. List three different multiples of 28 that are more than 100 and less than 200.

52. First, find the GCF of the numbers. Then factor the expressions using the GCF.

a. The GCF of 18 and 21 is ______

18 + 21 = ____ · ____ + ____ · ____ = ____ (____ + ____)

b. The GCF of 56 and 35 is ______

56 + 35 = ____ (____ + ____)

Fractions

53. Solve.

a. $\frac{4}{5} \div \frac{1}{5}$	b. $3\frac{1}{8} \div 1\frac{1}{2}$	c. $4 \div \frac{5}{7}$

54. Write a division sentence, and solve.

How many times does  go into ?

55. Write a real-life situation to match this fraction division: $1\frac{3}{4} \div 3 = \frac{7}{12}$

56. How many 3/4-cup servings can you get from 7 1/2 cups of coffee?

57. A rectangular room measures 12 1/2 feet by 15 1/3 feet. It is divided into three equal parts. Calculate the area of one of those parts.

58. The perimeter of a rectangular screen is 15 1/2 inches, and the ratio of its width to its height is 3:2. Find the width and height of the screen.

Integers

59. Compare the numbers, writing < or > in the box.

a. 0 ☐ −3 **b.** −2 ☐ −8

60. Write a comparison to match each situation (with < or >).

a. The temperature −7°C is warmer than −12°C.

b. Harry has $5. Emily owes $5.

61. Find the difference between the two temperatures.

a. −13°C and 10°C **b.** −9°C and −21°C

62. Write using mathematical symbols, and simplify (solve) if possible.

a. The opposite of 7. **b.** The absolute value of −6.

c. The absolute value of 5. **d.** The absolute value of the opposite of 6.

63. **a.** Plot the point (−5, 3).

b. Reflect the point in the x-axis .

c. Now, reflect the point you got in (b) in the y-axis.

d. Join the three points with line segments.
What is the area of the resulting triangle?

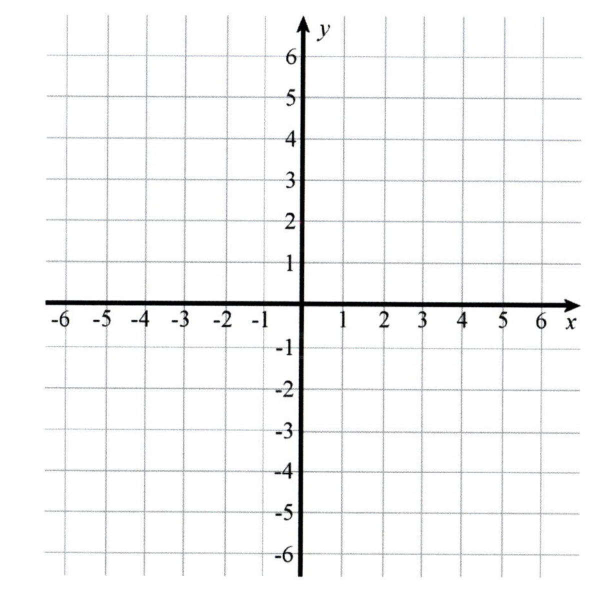

64. Draw a number line jump for each addition or subtraction sentence, and solve.

a. −2 + 5 = 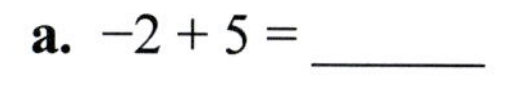

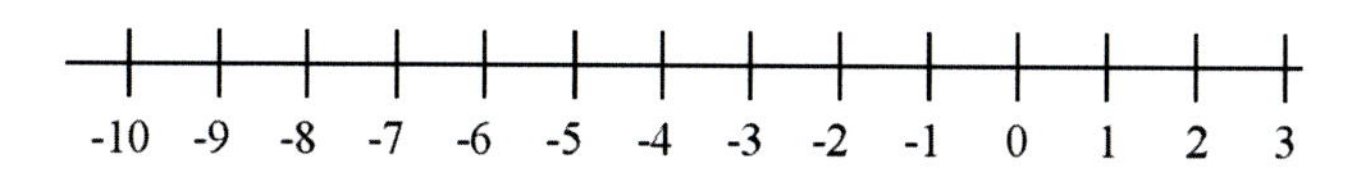

b. −2 − 4 =

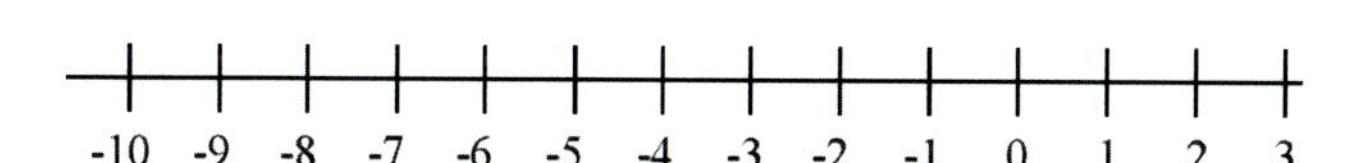

c. −1 − 5 = ______

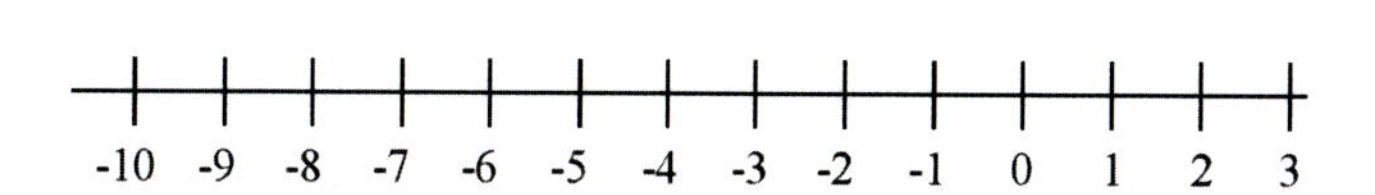

65. Write an addition or subtraction in the box to match each situation, and fill in the blanks.

a. Elijah has saved \$10. He wants to buy shoes for \$14.
That would make his money situation to be _____________.

b. A fish was swimming at the depth of 2 m. Then it sank 1 m.
Now he is at the depth of _________ m.

Geometry

66. Draw in the grid a right triangle with a base of 4 units and a height of 3 units.

Calculate its area.

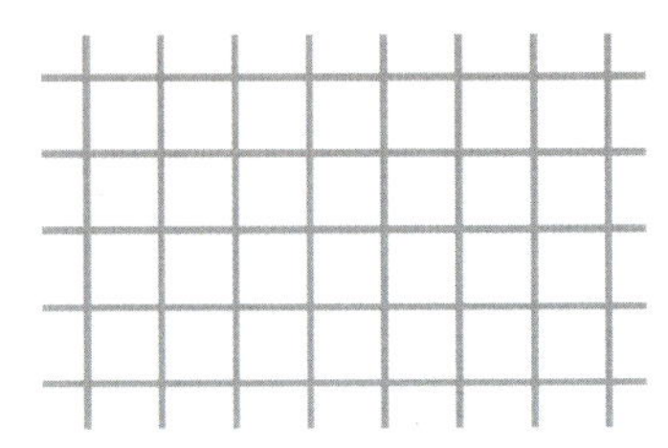

67. Draw in the grid a parallelogram with an area of 15 square units.

68. Find the area of this polygon, in square units.

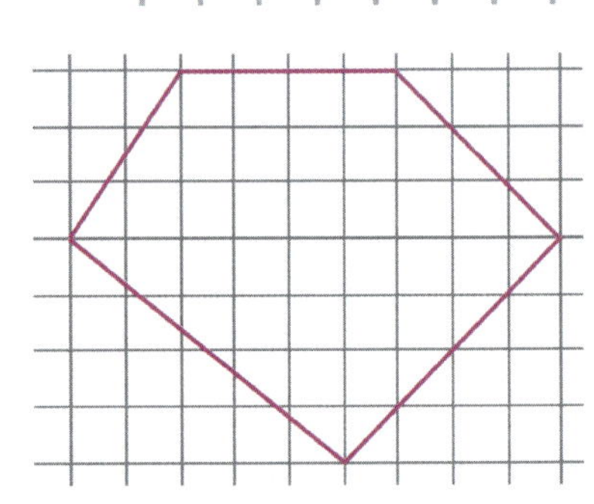

69. Draw a quadrilateral in the grid with vertices
$(-5, 5)$, $(-5, -3)$, $(2, -1)$, and $(2, 4)$.

What is the quadrilateral called?

Find its area.

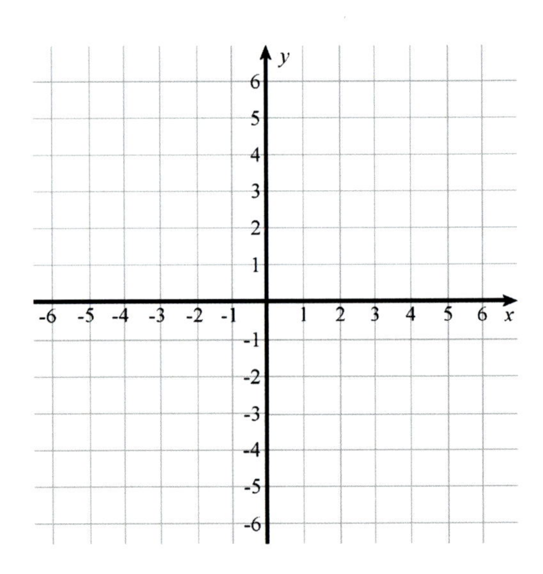

70. Name this solid. Draw a sketch of its net.

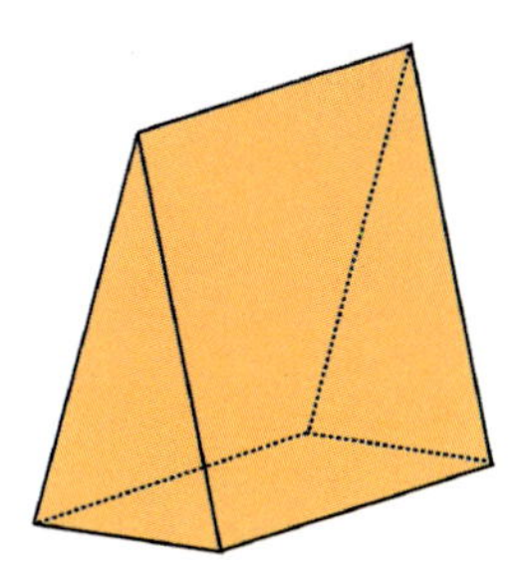

71. **a.** Name the solid that can be built from this net.

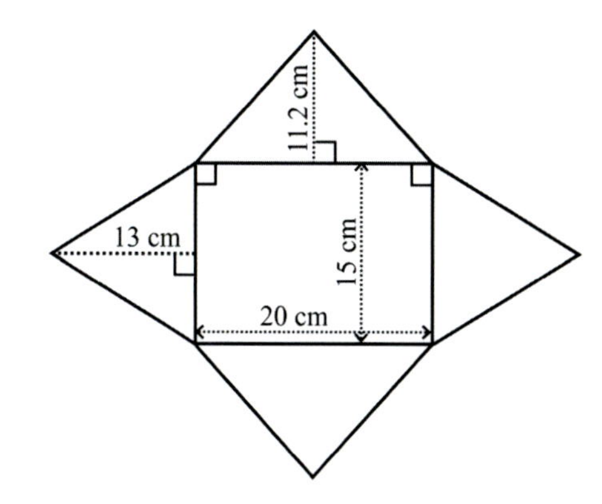

b. Calculate its surface area.

72. The edges of each little cube measure **1/2 cm**. What is the total volume of these figures, in cubic units?

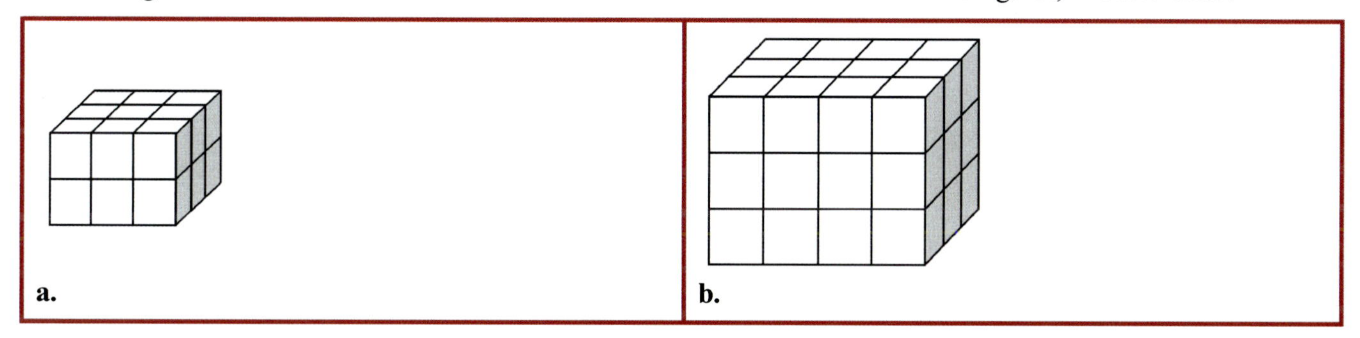

73. A box containing a construction toy measures 1 3/4 in by 8 1/2 in by 6 inches.

a. Calculate its volume.

b. How many of these boxes fit into a crate with the inside measurements of 1 ft by 1 ft by 1 ft?

74. **a.** Make a stem-and-leaf plot of this data.

55 59 61 62 64 65 65 68 69 70 72 74 77 83 89 94

(The ages of people in a senior chess club)

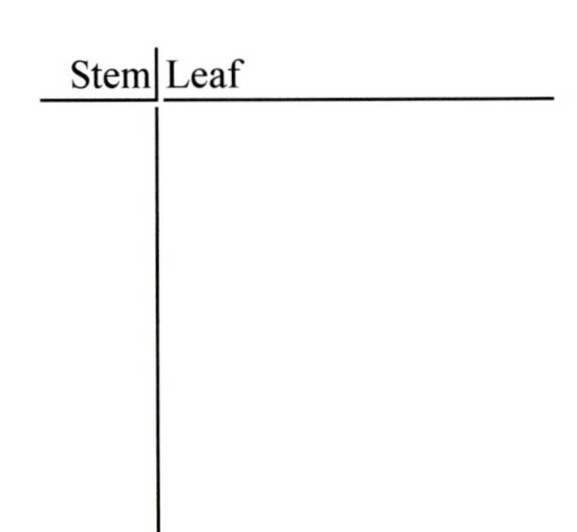

b. Find the median.

c. Find the interquartile range.

75. **a.** Describe the shape of this distribution.

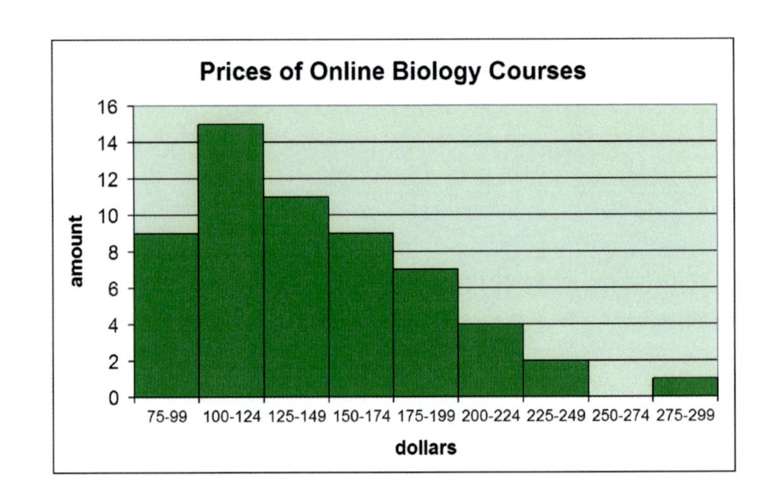

b. Which measure of center would be best to describe this distribution?

76. **a.** Create a dot plot from this data.

9 10 5 6 4 8 7 3 8 1 7 7 5 7 8 9 5 6 6 7

(points on a math quiz of a group of students)

b. Describe the shape of the distribution.

c. Describe the spread of the data.

d. Choose a measure of center to describe the data, and determine its value.

Math Mammoth Grade 6 Review Workbook Answers

The Basic Operations Review, p. 6

1. a. 83 R41 b. 6,735 R45

2. 23,391 ÷ 9 = 2,599 times

3. You will spend 365 × \$2.25 = \$821.25 in a year on phone calls.

4. 5,000 ÷ 46 = 108 R32. They will need 109 buses.

5. Multiply to estimate, and use 900 km, instead of 880 km. Since 6 × 900 km = 5,400 and 7 × 900 km = 6,300 km, it will take about 6 1/2 hours to travel 5,800 km.

6. \$15.90 ÷ 3 × 2 = \$10.60. Two boxes of tea bags cost \$10.60.

7. a. $5^4 = 625$ c. $30^2 = 900$ e. $2^6 = 64$
   b. $1^6 = 1$ d. $100^3 = 1{,}000{,}000$ f. $3^3 = 27$

8. a. Its area is 400 cm². (One side measures 20 cm.)
   b. Its volume is $(11 \text{ m})^3 = 1{,}331 \text{ m}^3$.

9. a. 25^3 gives us the volume of a cube with an edge length of 25 units.
   b. 3×9^2 gives us the area of 3 squares with a side length of 9 units.

10. a. 200,309 b. 28,031,000

11. a. $707{,}000 < 7{,}000{,}000 < 10^7$
    b. $5 \times 10^4 < 4 \times 10^5 < 450{,}000$

12. a. 149,601 ≈ 150,000 b. 2,999,307 ≈ 3,000,000
    c. 597,104,865 ≈ 597,000,000 d. 559,998,000 ≈ 560,000,000

The Basic Operations Test, p. 8

1. 10,540 R28

2. 1.909

3. \$937.50 ÷ 75 = \$12.50. One flashlight costs \$12.50.

4. It will take you 43 sec/page × 234 pages = 10,062 seconds = 167 minutes and 42 seconds (almost 3 hours).

5. a. $3^3 = 27$ b. $1^{10} = 1$ c. $50^2 = 2{,}500$ d. $10^5 = 100{,}000$

6. The length of each side is 56 cm ÷ 4 = 14 cm. So the area is 14 cm × 14 cm = 196 cm².

7. a. 504,300,000
   b. 1,600,020,100

8. a. $5 \times 10^5 + 6 \times 10^4$
   b. $9 \times 10^6 + 1 \times 10^5 + 8 \times 10^3$

9. a. 3,000,000 b. 480,000,000 c. 20,000,000

Expressions and Equations Review, p. 10

1. a. $(6 - x)^2$ b. $\frac{5}{x + 6}$ c. $3(5 - p)$

2. a. 113 b. 200
   c. 9 d. 560

3. a. 28 b. 91

4. a. $p \div 3$
   b. \$3 + 6c OR 6c + \$3

5.

$2x + 17$	$8 = 8$	$y < 5$	$4x - 3 = 8$	$\frac{4}{5}x - 16$	$4x + y^2 \geq 9$	$M = \frac{44 - x}{5}$
expression	equation	inequality	equation	expression	inequality	equation

6.

a. $3t + 3$	b. $5d$
c. x^3	d. $6x - 6$
e. $16z^3$	f. $14x^2 + 5$

7.

a. $A = 9s^2$ $P = 12s$	b. $A = 6x^2$ $P = 10x$

8. a. $6x + 21$
   b. $72b + 40$

9.

a. $5x + 10 = 5(x + 2)$	b. $6y + 10 = 2(3y + 5)$
c. $24b + 4 = 4(6b + 1)$	d. $25w + 40 = 5(5w + 8)$

10. a. 112 b. 116 c. 72
    d. 60 4/13 e. 328 f. 70

11. a. $25q = 1675$. $q = 67$. There are 67 quarters.

    b. $2(x + 21) = 128$ OR $x + x + 21 + 21 = 128$ OR $2x + 21 + 21 = 128$. $x = 43$.
    The other side is 43 meters long.

12. The temperature is 77 degrees Fahrenheit.

13. a. $x < 57$ b. $x \geq 30$

14. a. 23, 30, 55, 44

    b. 2, 4, 6

15. $y = x + 3$

x	1	2	3	4	5	6
y	4	5	6	7	8	9

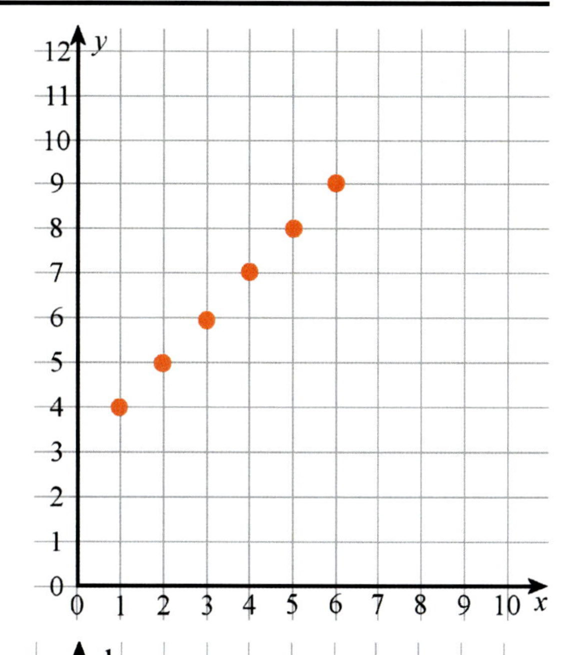

16. $y = x + 3$

t (hours)	0	1	2	3	4	5	6
d (miles)	0	70	140	210	280	350	420

c. $70t = d$
d. t

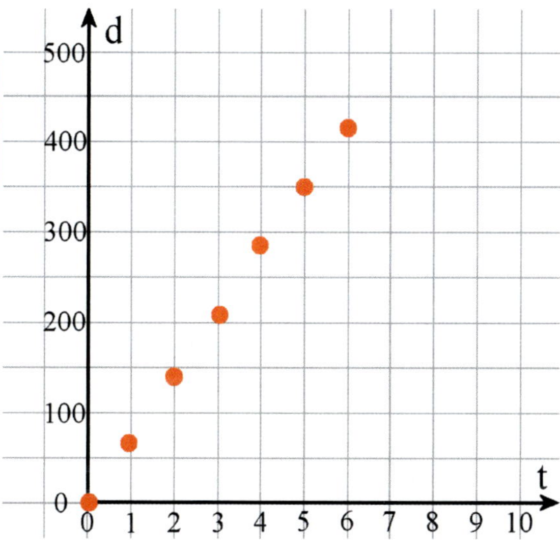

Expressions and Equations Test, p. 14

1. a. $x^2/7$ b. $(5 - y)^3$ c. $3(2s - 5)$

2. a. 20 b. 60 c. 32

3. a. 20 b. 35 c. 7 d. 39

4. $p + 3t$

5. a. a^4 b. $4a$ c. $10x^2$ d. $6d + 7$

6. a. $5(x + 6) = 5x + 5 \cdot 6 = 5x + 30$
   b. $2(9 + 5y) = 2 \cdot 9 + 10y = 18 + 10y$

7. a. $x = 144 \div 6 = 24$
   b. $y = 134 - 78 = 56$
   c. $x = 3 \cdot 16 = 48$

8. Let one side be denoted by s. The equation is $4s = 164$. Solution: $s = 41$

Expressions and Equations Test, cont.

9.

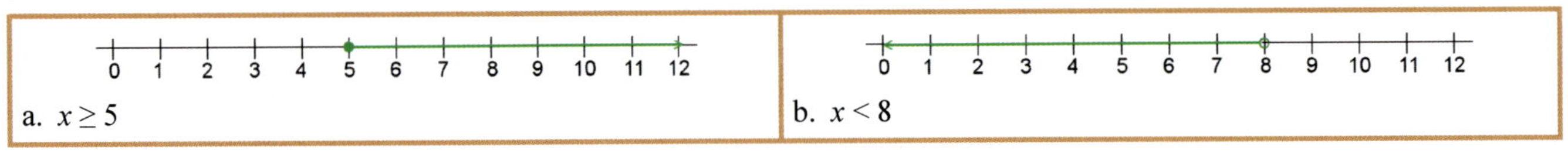

10. 15, 16

11. $y = 9 - x$.

x	0	1	2	3	4	5	6	7	8	9
y	9	8	7	6	5	4	3	2	1	0

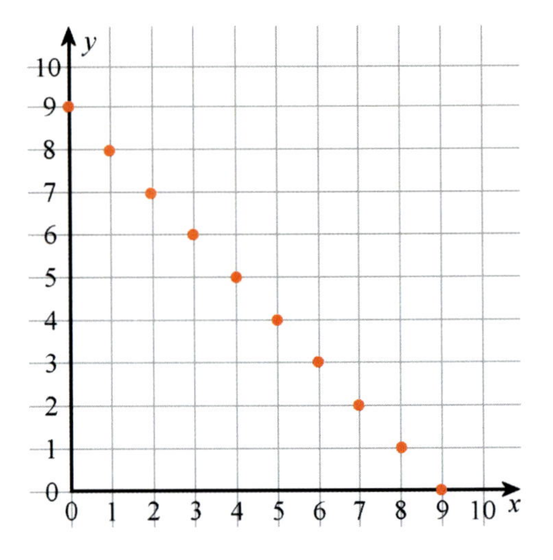

Mixed Review 1, p. 16

1. a. 100,000 b. 81 c. 400,000

2. 10^7 is ten million.

3. a. A = 3 km · 3 km = 9km^2
   b. V = 2 in · 2 in · 2 in = 8 in^3

4. a. 4 × 9 cm = 36 cm. The perimeter is 36 cm.
   b. 48 m ÷ 4 = 12 m. The area is 12 m × 12 m = 144 m^2.

5. a. 5,051,000,027,000
   b. 21,650,099,000,056

6. 6,002,001

7. a. $5 \cdot 10^4 + 4 \cdot 10^3$
   b. $2 \cdot 10^6 + 9 \cdot 10^4 + 3 \cdot 10^1$

8. a. $5s \div 8$ b. $7(x + 8)$ c. $8 - y$ d. $(x - 8)^2$

9. Estimations may vary.

a. 591 · 57,200 Estimation: 600 · 57,000 = 34,200,000 OR 600 · 60,000 = 36,000,000 Exact: 33,805,200 Error of estimation: 394,800 OR 2,194,800	b. 435,212 + 9,319,290 Estimation: 435,000 + 9,320,000 = 9,755,000 Exact: 9,754,502 Error of estimation: 498

10.

a. $4 \cdot 50 + \dfrac{310}{2} = 355$	b. $\dfrac{4{,}800}{60} - (70 - 20) = 30$

11. The price is reduced by 2/10. So \$120 represents the remaining 8/10 of the original cost.
    We can find 1/10 of the price by dividing the remaining price by 8: \$120 ÷ 8 = \$15.
    Now, \$15 is 1/10 of the original cost; therefore the original price was 10 · \$15 = \$150.

12. 7123 R73

Mixed Review 2, p. 18

1. a.

Distance	4 km	12 km	20 km	24 km	132 km	216 km
Tim	10 min	30 min	50 min	1 hour	5 1/2 hours	9 hours

b. To travel 360 kilometers will take 360 km ÷ 24 km/hr = 15 hours.

2. a. Answers may vary. Check the student's answers. The simplest estimate is to round 1,091 down to 1,000: 234 · 1,000 = <u>234,000</u>.
(You might compensate for rounding one number down by rounding the other one up: 240 · 1,000 = 240,000.)

b. 234 · 1,091 = 255,294

c. Answers may vary. Check the student's answers. The simplest estimate is to round 1.091 down to 1: 2.34 · 1 = <u>2.34</u>.
(You might compensate for rounding one number down by rounding the other one up: 2.4 · 1 = 2.4.)

d. Adjusting the decimal points in our estimate from part (a), we can see that 2.34 · 1 is 2.34, so we also know that 2.34 · 1.091 is about that same amount, so, based on part (b), the exact answer must be 2.55294.

Enrichment (optional): Exercise #2 is a fascinating problem because it lends itself so well to what is called an "iterative" (*i.e.*, "in steps") method. Although the simplest estimate simply rounds 1,091 down to 1,000, that estimate will be too small because of rounding down. We could get a little closer by recognizing that 91 is almost 100. (We are rounding 91 up, so this estimate will be too big.) Since 234 · 100 = 23,400, a closer estimate would be 234 · 1,100 = 234 · 1,000 + 234 · 100 = 234,000 + 23,400 = 257,400. The exact answer is somewhere between 234,000 and 257,400, but a lot (91/100) closer to 257,400. We can even keep going, subtracting 234 · 10 = 2,340 to get 234 · 1,090 = 257,400 − 2,340 ≈ 255,000, which gets us accuracy to the nearest thousand. In other words, 234 · 1,091 = 234 · 1,000 + 234 · 100 − 234 · 10 + 234 · 1, so we can transform the multiplication problem into a simpler addition/subtraction problem and estimate the answer to whatever place value we want.

3. a. $56 - y = 17$ (*Minuend − subtrahend = difference.*) Solution: $y = 56 - 17 = 39$.
b. $x \div 15 = 60$ (*Dividend ÷ divisor = quotient.*) Solution: $x = 60 \cdot 15 = 900$.

4. a. 600 ÷ 6 + 36 ÷ 6 = 100 + 6 = 106
b. 800 ÷ 4 + 24 ÷ 4 = 200 + 6 = 206
c. 5,600 ÷ 7 + 7 ÷ 7 = 800 + 1 = 801
d. 1,200 ÷ 12 + 24 ÷ 12 = 100 + 2 = 102

5. a. 100 − (100 ÷ 4) · 2 = 100 − (25 · 2) = 100 − 50 = <u>50</u>
b. $3^3 \div (4 + 5) = 27 \div 9 =$ <u>3</u>
c. $(2 + 6)^2 - (25 - 5) = 8^2 - 20 = 64 - 20 =$ <u>44</u>
d. (144 + 9)/(5 · 3) = 153/15 = <u>10 1/5</u>

6. a. 3 · 5 − 12 = 15 − 12 = 3
b. 24/3 + 4 = 8 + 4 = 12

7. The perimeter is the sum of the lengths of the sides, or $l + w + l + w$, so the correct answer is <u>(b) $2l + 2w$</u>.
(Answer (a) gives the rectangle's area. Answer (c) gives the aspect ratio. Answers (d) and (e) give partial perimeters.)

8. a. $A = (11 \text{ cm})^2 = 121 \text{ cm}^2$
b. $V = (4 \text{ ft})^3 = 64 \text{ ft}^3$

9. To find the length of one side of the square divide 64 cm by 4. Then the area would be $A = (16 \text{ cm})^2 = 256 \text{ cm}^2$.

10. $48.60/3 · 2 = $16.20 · 2 = $32.40. The total cost is $48.60 + $32.40 = $81. Her change is $19.00.

11. a. 15,711 b. 0.533 c. 0.043

Decimals Review, p. 20

1. a. three ten-thousandths = 0.0003
   b. 39234 hundred-thousandths = 0.39234
   c. 4 millionths = 0.000004
   d. 2 and 5 thousandths = 2.005

2. a. 0.00039 = 39/100,000 b. 0.0391 = 391/10,000 c. 4.0032 = 40,032/10,000

3. a. 0.75 b. 1.4 c. 0.85 d. 0.44

4.

Organism	Size (fraction)	Size (micrometers)	Size (decimal)
amoeba proteus	$\frac{600}{1,000,000}$ meters	600 micrometers	0.0006 m
protozoa	from $\frac{10}{1,000,000}$ to $\frac{50}{1,000,000}$ m	from 10 to 50 micrometers	from 0.00001 to 0.00005 m
bacteria	from $\frac{1}{1,000,000}$ to $\frac{5}{1,000,000}$ m	from 1 to 5 micrometers	from 0.000001 to 0.000005 m

5. a. 0.000526 $<$ 0.0062 $<$ 0.0256 b. 0.000007 $<$ 0.00008 $<$ 0.000087

6.

	0.37182	0.04828384	0.39627	0.099568
To the nearest hundredth:	0.37	0.05	0.40	0.10
To the nearest ten-thousandth:	0.3718	0.0483	0.3963	0.0996

7. a. 0.024 b. 0.75 c. 3.043

8. a. 2.1 − 1.09342 = 1.00658
   b. 17 + 93.1 + 0.0483 = 110.1483

9. a. 0.1 + 0.04 = 0.14 b. 0.01 + 0.04 = 0.05 c. 0.0001 + 0.04 = 0.0401

10.

a. 0.48 ÷ 6 = 0.08 6 × 0.08 = 0.48	b. 1.5 ÷ 0.3 = 5 0.3 × 5 = 1.5	c. 0.056 ÷ 0.008 = 7 0.008 × 7 = 0.056

11.

a. 3 × 0.006 = 0.018	b. 0.2 × 0.6 = 0.12	c. 0.9 × 0.0007 = 0.00063

12. 327 × 4 is 1,308. In the calculation 32.7 × 0.004, the decimals have one and three decimal digits, or four decimal digits in total. So we take 1,308 and make it have four decimal digits, so it becomes 0.1308.

13. a. Estimate: 9 × 0.06 = 0.54.
    b. Exact: 8.9 × 0.061 = <u>0.5429</u>

14. a. 0.03 b. 0.12 c. 0.05

15. a. $p = 225$ b. $x = 173.33$ c. $y = 0.324$

16. There is 4 m − (7 × 0.56 m) = 4 m − 3.92 m = <u>0.08 m</u> left.

17.

a. $10^6 \times 21.7 = 21{,}700{,}000$	b. 100 × 0.00456 = 0.456
c. 2.3912 ÷ 1,000 = 0.0023912	d. $324 \div 10^5 = 0.00324$
e. $10^5 \times 0.003938 = 393.8$	f. $0.7 \div 10^4 = 0.00007$

18. $\frac{a}{b} + 1 = 3.585$

19. a. 14.1 b. 0.007 c. 0.444 d. 0.455

Decimals Review, cont.

20.

Prefix	Meaning	Units - length	Units - mass	Units - volume
centi-	hundredth = 0.01	centimeter (cm)	centigram (cg)	centiliter (cl)
deci-	tenth = 0.1	decimeter (dm)	decigram (dg)	deciliter (dl)
deca-	ten = 10	decameter (dam)	decagram (dag)	decaliter (dal)
hecto-	hundred = 100	hectometer (hm)	hectogram (hg)	hectoliter (hl)

21. a. 34 dl = 3.4 L b. 89 cg = 0.89 g c. 16 kl = 16,000 L

22. a. 2.7 L = 27 dl = 270 cl = 2700 ml
 b. 5,600 m = 5.6 km = 56,000 dm = 560,000 cm
 c. 676 g = 6,760 dg = 67,600 cg = 676,000 mg

23. The total capacity is 6 × 0.35 L + 2 × 2 L + 3 × 0.9 L = 2.1 L + 4 L + 2.7 L = 8.8 L.

24. a. 56 oz = 3.5 lb c. 2.7 gal = 10.8 qt e. 0.48 mi = 2,534.4 ft
 b. 134 in = 11.17 ft d. 0.391 lb = 6.26 oz f. 2.45 ft = 2 ft 5.4 in

25. The ribbons will measure 500 ft ÷ 230 = 2.17 ft = 2 ft 2 in.

26. The 40 yards of rope costs $15.99 / 40 yd = $0.40 per yard, and the 100 meters costs $40 ÷ 100 = $0.40 per meter. A meter is longer than a yard so the 100 meters is the better deal.

27. Since 3.2 + 3.1 + 3.4 + 3.1 + 3.5 + 2.9 + 2.7 + 2.7 + 3.0 + 3.0 + 3.1 + 3.4 + 3.2 + 2.8 + 2.8 + 2.9 + 3.6 + 3.4 + 2.9 + 3.4 + 3.1 = 65.2, the average length of the 21 tadpoles was 65.2 / 21 = 3.1 cm.
 (We give the answer to tenths because that was the accuracy of all of the data.)

Decimals Test, p. 24

1. a. 0.005 b. 0.00382 c. 1.003658 d. 0.0094 e. 0.65 f. 8.08

2. a. 20,045/10,000 b. 912/1,000,000 c. 749,038/100,000

3. The student may give the answer either as a decimal or as a fraction — both are correct.
 a. 0.2 + 0.005 = 0.205 OR 205/1000
 b. 0.07 + 0.03 = 0.10 = 0.1 OR 1/10
 c. 2.022 + 0.033 = 2.055 or 2 55/1000

4. a. 2.31 × 0.04 = 0.0924
 b. 3.38758 ÷ 7 + 0.045 = 0.52894

5.

Round to ...	0.0882717	0.489932	1.299959
... the nearest thousandth	0.088	0.490	1.300
... the nearest hundred-thousandth	0.08827	0.48993	1.29996

6. a. 0.24 ÷ 3 = 0.08 b. 5.4 ÷ 0.6 = 9 c. 0.081 ÷ 0.009 = 9
 d. 2 × 0.05 = 0.1 e. 8 × 0.009 = 0.072 f. 11 × 0.0005 = 0.0055

7. The area of the parcel is 50.5 m × 27.6 m = 1,393.8 m^2, so each fourth has an area of 1,393.8 m^2 ÷ 4 = 348.45 m^2.

8. a. 20 b. 0.47 c. 1097000 d. 0.006 e. 0.001245 f. 0.00324

9. a. 0.04 ÷ 4 = 0.01 b. 0.04 ÷ 0.04 = 1 c. 0.04 ÷ 10 = 0.004

10. a. 56 mm = 0.056 m b. 9 km = 9,000 m c. 9 cg = 0.09 g d. 16 dl = 1.6 L

11. a. 2.7 km = 2,700 m = 270,000 cm = 2,700,000 mm
 b. 5,600 ml = 560 cl = 56 dl = 5.6 L
 c. 0.6 g = 6 dg = 60 cg = 600 mg

Decimals Test, cont.

12. Since 7 pounds 6 ounces = 7 6/16 lbs = 7.375 lbs, 7.4 pounds is heavier than 7 pounds 6 ounces by 7.4 lbs − 7.375 lbs = 0.025 lbs (= 1/40 lb = 4/10 ounce).

13. The pint-sized bottle is $7/16 oz = 43.75¢ per ounce, and the 24-ounce size is $12/24 oz = 50¢ per ounce. The pint-sized bottle of honey is the better deal. You can also solve this by figuring out the price for 48 ounces: Three pint-sized bottles (48 oz in total) cost $21, whereas two of the 24-ounce bottles cost $24. The pint-sized bottles are the better deal.

14. a. $5.36 \div 0.2 = 26.8$ b. $1.6 \div 0.05 = 32$

c. $22.9 \div 7 = 3.271$ d. $\frac{8}{9} = 0.889$

Mixed Review 3, p. 27

1. Ten to the power of eight is equal to a hundred million. $10^8 = 100{,}000{,}000$.

2. $3 \times 10^6 + 5 \times 10^5 + 4 \times 10^2 + 8 \times 10^1$

3.

a. $213 \cdot 5{,}829$ Estimation: $200 \cdot 5{,}900 = 1{,}180{,}000$ Exact: 1,241,577 Error of estimation: 61,577	b. $435{,}212 \div 993$ Estimation: $435{,}000 \div 1{,}000 = 435$ Exact: 438.28 Error of estimation: 3.28

4.

c	$c + \frac{2c}{5}$
15	$10 + \frac{2 \cdot 15}{5} = 15 + 6 = 21$

c	$c + \frac{2c}{5}$
20	$10 + \frac{2 \cdot 20}{5} = 20 + 8 = 28$
25	$10 + \frac{2 \cdot 25}{5} = 25 + 10 = 35$

5. $5.1 \div 3 \times 10 = 17$. A full gas tank holds 17 gallons.

6. $98 ÷ 7 × 6 + $98 = $182. The total cost for both printers was $182.

7.

a. $\frac{15 + 150}{5} = 33$	b. $\frac{5}{15 + 5} = \frac{1}{4}$	c. $\frac{380 + 10}{12 - 9} = 130$

8. a. $(t - 1)^2$ b. $9 - x$

c. $7 + S$ d. $8(4 + x + 2)$

e. $x^2 \div (x + 1)$

9. a. $24 - 11 = 13$

b. $(3/5) \cdot 7 = 21/5 = 4\ 1/510.$

10.

a. x^5	b. $2p + 2$
c. $10 \cdot x^3$	d. $8z$
e. $3f + 2x$	f. $3s + 10$

11. a. $t \geq 18$ b. $p \leq \$40$ c. $a > 12$

12. 3, 4, 5, 6

Mixed Review 3, cont.

13. a. $3(5x + 6) = 15x + 18$
    b. $2(8x + 2 + y) = 16x + 4 + 2y$

14.

a.	b.	c.
$x + 78 = 412$ $x = 412 - 78$ $x = 334$	$\frac{x}{9} = 600$ $x = 9 \cdot 600$ $x = 5{,}400$	$y - 5 = 12 + 18$ $y = 30 + 5$ $y = 35$

Mixed Review 4, p. 29

1. One parent paid \$103.74 and the other three parents paid \$34.58 each.

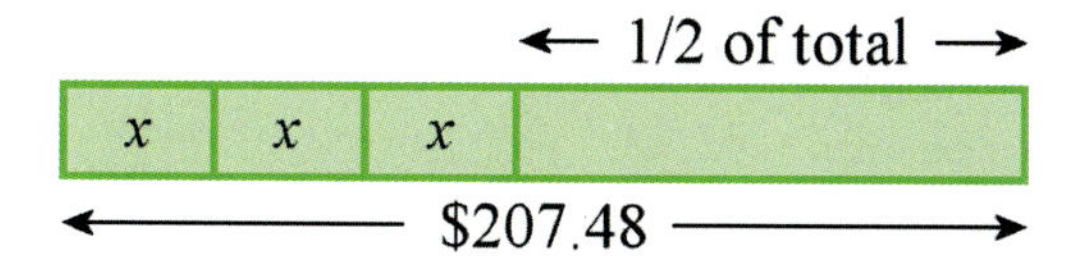

2. a. 329,000,300 b. 1,050,003

3. a. 5,700,000 b. 219,997,000 c. 83,000,000 d. 3,999,990,000

4.

Variable	Expression $\frac{x^2}{3}$	Value
$x = 2$	$\frac{2^2}{3} = \frac{4}{3}$	$1\frac{1}{3}$

Variable	Expression $\frac{x^2}{3}$	Value
$x = 3$	$\frac{3^2}{3}$	3
$x = 5$	$\frac{5^2}{3} = \frac{25}{3}$	$8\frac{1}{3}$

5. a. $(12 + 56) \div 4$ or $\frac{12 + 56}{4} = 68 \div 4 = 17$.
   b. $8 \div 4^3$ or $\frac{8}{4^3} = \frac{1}{8}$.

6. a. $8c^4$ b. $5c + 8$
   c. $t + 3$ d. $13x^2 + 13$

7. a. Her friend got $(1/3)m$ or $m/3$.
   b. Fanny is $s - 6$.
   c. Sadie will be $s + 5$.
   d. Fanny will be $s - 1$.

8.

a.	b.	c.
$7x + 2x = 54$ $9x = 54$ $x = 54 \div 9$ $x = 6$	$8r - 3r = 40$ $5r = 40$ $r = 40 \div 5$ $r = 8$	$t \div 50 = 5 + 6$ $t \div 50 = 11$ $t = 50 \cdot 11$ $t = 550$
d.	**e.**	**f.**
$w - 88 = 20 \cdot 60$ $w - 88 = 1{,}200$ $w = 1{,}200 + 88$ $w = 1{,}288$	$2x - 6 = 16$ $2x = 16 + 6$ $x = 22 \div 2$ $x = 11$	$8x + 17 = 81$ $8x = 81 - 17$ $x = 64 \div 8$ $x = 8$

Mixed Review 4, cont.

9.

a. $16y + 12 = 4(4y + 3)$	b. $9x + 9 = 9(x + 1)$
c. $54c + 24 = 6(9c + 4)$	d. $15a + 45 = 15(a + 3)$

10. a. $x = 7 - 4.5039 = 2.4961$
 b. $x = 0.938208 - 0.047 = 0.891208$
 c. $x = 6.0184 \div 2 = 3.0092$

Ratios Review, p. 31

1. a. $\frac{4}{3} = \frac{20}{15}$	b. 6:7 = 18:21	c. 4 to 30 = 2 to 15	d. $\frac{7}{3} = \frac{28}{12}$

2. a. $\frac{15}{35} = \frac{3}{7}$	b. $\frac{6}{16} = \frac{3}{8}$	c. 33:30 = 11:10	d. 9:12 = 3:4

3. a.

b. 2/3 heart for **1** triangle

3/2 triangles for **1** heart

4.

Miles	50	100	150	200	250
Hours	1	2	3	4	5

Miles	300	350	400	450	500
Hours	6	7	8	9	10

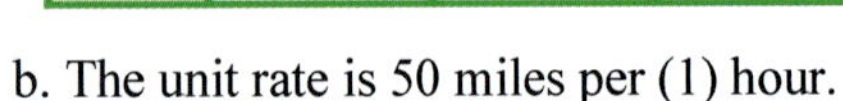

b. The unit rate is 50 miles per (1) hour.
c. The car would go 375 miles in 7 1/2 hours.
d. It would take 4 1/2 hours to travel 225 miles.

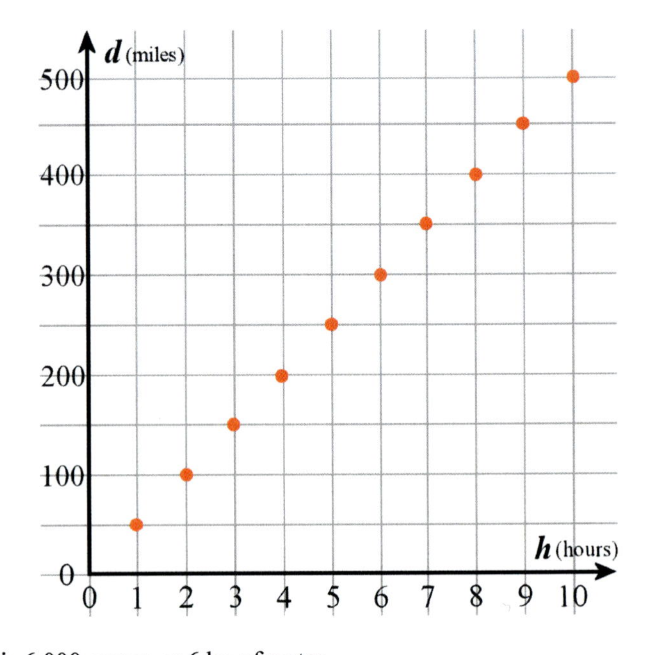

5. a. 20 g of salt : 1200 g of water = 1 : 60
 b. For 100 grams of salt you need 60 times as much water, which is 6,000 grams or 6 kg of water.

6. In the ratio 11:12, the 12 parts represent Dad's age (he's older) and the 11 parts represent Mom's. The 3-year difference in their ages is the one part difference in the ratio. Therefore, Dad is $3 \times 12 = 36$ years old and Mom is 33 years old.

7. A bean plant is 3/5 as tall as a tomato plant. The tomato plant is 20 cm taller than the bean plant.
 a. The ratio of the bean plant's height to the tomato plant's height is 3:5
 b. The tomato plant is 20 cm or 2 parts taller than the bean plant so each part is 10 cm.
 The bean plant is $3 \times 10 = 30$ cm tall. The tomato plant is $5 \times 10 = 50$ cm tall.

8. $63 \text{ cm} \div 9 \times 16 = 112$ cm. The television screen is 112 cm wide.

Ratios Review, cont.

9. a. \$20/12 kg × 5 kg = \$8.33. Or, you can first find the unit rate. 20 dollars : 12 kg = 20/12 dollars per kg = 1 8/12 dollars per kg = 1 2/3 dollars per kg. Then, multiply that by 5 to get the price for 5 kg.
   b. The unit rate is 20 dollars : 12 kg = 20/12 dollars per kg = 1 8/12 dollars per kg = 1 2/3 dollars per kg = \$1.67 per kg.

10.

a. $134 \text{ lb} = 134 \text{ lb} \cdot \dfrac{1 \text{ kg}}{2.2 \text{ lb}} = \dfrac{134 \text{ kg}}{2.2} \approx 60.91 \text{ kg}$
b. $156 \text{ cm} = 156 \text{ cm} \cdot \dfrac{1 \text{ in.}}{2.54 \text{ cm}} \cdot \dfrac{1 \text{ ft}}{12 \text{ in.}} = \dfrac{156 \text{ ft}}{2.54 \cdot 12} \approx 5.12 \text{ ft}$

Ratios Test, p. 33

1. a. 3/5 = 18/30   b. 2:3 = 18: 27   c. 10 to 45 = 2 to 9   d. 12:30 = 2 to 5

2. a.

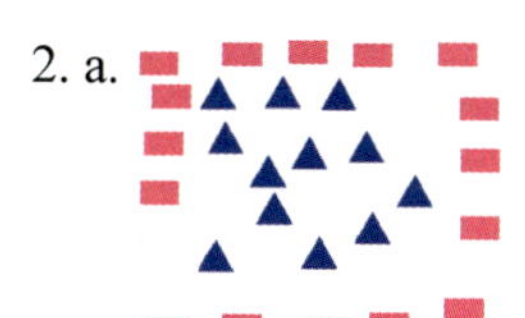

   b. 1 1/3 rectangles for **1** triangle (the ratio of rectangles to triangles is 16:12 = 4:3 = 4/3 : 1)

   3/4 triangles for **1** rectangle (the ratio of triangles to rectangles is 12:16 = 3:4 = 3/4 : 1)

3. a. $\dfrac{4 \text{ L}}{10 \text{ m}^2} = \dfrac{2 \text{ L}}{5 \text{ m}^2} = \dfrac{10 \text{ L}}{25 \text{ m}^2}$ b. $\dfrac{\$9}{6 \text{ min}} = \dfrac{\$3}{2 \text{ min}} = \dfrac{\$15}{10 \text{ min}} = \dfrac{\$90}{1 \text{ hour}}$

4. a. The unit rate is 0.1 m per 1 minute.
   b. 17 min · 0.1 m/min = 1.7 meters.

5. a. $\dfrac{14 \text{ downloads}}{\$2.10} = \dfrac{2 \text{ downloads}}{\$0.30} = \dfrac{1 \text{ download}}{\$0.15} = \dfrac{3 \text{ downloads}}{\$0.45}$, so three song downloads would cost 45¢.

6. The ratio is 8:5 and the shorter side length of the rectangle is 15 cm, so each "part" in the ratio corresponds to 3 cm.
   a. The rectangle's length is 8 · 3 cm = 24 cm.
   b. Its area is 15 cm · 24 cm = 360 cm^2.

7. The 1 + 7 = 8 parts make a total of 4 L, so each part is 4 L ÷ 8 = ½ L. You need ½ L of juice concentrate and 7 × ½ L = 3 ½ L of water.

8. a. 35 gal per 7 mi = 5 gal per 1 mi.
   b. The plane can fly 100 miles with 500 gallons of fuel.
   c. The plane will need 750 gallons of fuel to travel 150 miles.

9. There are 3 + 5 = 8 parts, so each part consists of 1,200 ÷ 8 = 150 inserts. Anita folded 3 · 150 = <u>450</u> inserts and Michael folded 5 · 150 = <u>750</u> inserts.

10.

a. $60 \text{ cm} = 60 \text{ cm} \cdot \dfrac{1 \text{ in}}{2.54 \text{ cm}} = \dfrac{60 \text{ in}}{2.54} \approx 23.6 \text{ in}$
b. $4.5 \text{ ft} = 4.5 \text{ ft} \cdot \dfrac{30.48 \text{ cm}}{1 \text{ ft}} = 4.5 \cdot 30.48 \text{ cm} = 137.16 \text{ cm} \approx 137 \text{ cm}$

Mixed Review 5, p. 35

1. 4,958/13 = 381 R5

2. 43 ÷ 9 = 4.778

3. a. 51,99̲9,601 ≈ 52,000,000
   b. 109,9̲99,339 ≈ 110,000,000

4.

a. $3 \times 0.25 = 0.75$ $4 \times 0.025 = 0.1$	b. $8 \times 0.08 = 0.64$ $100 \times 0.0008 = 0.08$	c. $1 \div 0.05 = 20$ $4 \div 0.05 = 80$	d. $0.99 \div 11 = 0.09$ $0.06 \div 0.001 = 60$

5.

a. $10^5 \times 3.07 = 307000$	b. $10^4 \times 0.00078 = 7.8$
c. $12.7 \div 10^3 = 0.0127$	d. $5{,}600 \div 10^5 = 0.056$

6. The length of each side is $4y$.

7.

Expression	the terms in it	coefficient(s)	Constants
$2a + 3b$	$2a$ and $3b$	2 and 3	
$10s$	$10s$	10	
$11x + 5$	$11x$ *and 5*	11	5
$8x^2 + 9x + 10$	$8x^2$ and $9x$ and 10	8 and 9	10
$\frac{1}{6}p$	$\frac{1}{6}p$	$\frac{1}{6}$	

8. a.

X	1	2	3	4	5	6	7	8	9
Y	9	8	7	6	5	4	3	2	1

b. $X + Y = 10$ or $Y = 10 - X$.
c. In this problem, one could choose either one of the variables to be plotted on the horizontal axis. The most likely choice is X, though.

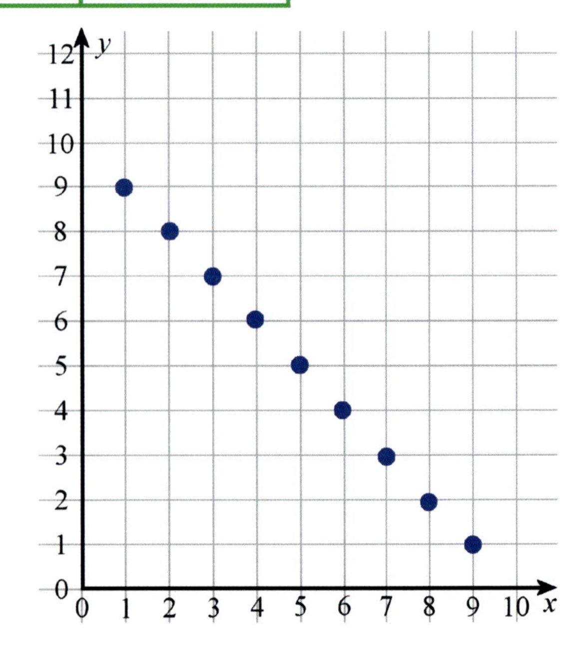

9. a. $5n$ b. $67 - y$ c. $\frac{8}{10}p$

10. a. $675.5 \div 0.3 = 2{,}251.667$

b. $\frac{2}{7} = 0.286$

Mixed Review 6, p. 37

1. a. Equation: $x \div 11 = 12$ or $\frac{x}{11} = 12$. Solution: $x = 11 \cdot 12 = 132$.

   b. Equation: $3 + 5 + x = 105$. Solution: $x = 105 - 5 - 3 = 97$.

2.

a.	b.	c.
$x \div 6 = 40 + 50$	$1{,}000 - x = 40 \cdot 6$	$8x + 2x = 15 \cdot 6$
$x \div 6 = 90$	$1{,}000 - x = 240$	$10x = 90$
$x = 90 \cdot 6$	$1{,}000 = 240 + x$	$x = 90 \div 10$
$x = 540$	$1000 - 240 = x$	$x = 9$
	$760 = x$	

3. a. The average price is the sum of the values divided by the number of items:
   ($3.89 + $3.99 + $4.45 + $3.79 + $4.10 + $4.19 + $4.02) / 7 = $28.43 / 7 = $4.06
   b. She saves $445 − $379 = $66.

4.

a. $10 \cdot 0.009 = 0.09$ $0.5 \cdot 0.6 = 0.3$	b. $40 \cdot 0.08 = 3.2$ $1{,}000 \cdot 1.2 = 1200$	c. $0.1 \cdot 0.2 \cdot 0.3 = 0.006$ $0.11 \cdot 0.02 = 0.0022$
d. $10 \div 0.2 = 50$ $0.6 \div 0.2 = 3$	e. $0.075 \div 0.025 = 3$ $0.3 \div 0.02 = 15$	f. $2.36 \div 2 = 1.18$ $0.0045 \div 5 = 0.0009$

5.

a. 6 kg = <u>6,000</u> g 5 dl = <u>0.5</u> L 5 mm = <u>0.005</u> m	b. 7 dam = <u>70</u> m 5 hl = <u>500</u> L 30 cg = <u>0.3</u> g	c. 7 kl = <u>7,000</u> L 50 mg = <u>0.05</u> g 8 cm = <u>0.08</u> m

6. a. Tim's weight compared to the grasshopper's is 45,000 g : 3 g, so <u>Tim weighs 15,000 times more</u>.
   b. You could easily carry the weight of a thousand grasshoppers because it would be only $1000 \cdot 3$ g = 3 kg (about 6 ½ lb).

7. According to the graphic, 5/6 of Elaine's after-tax salary (in brown) was $1,000, so each sixth (brown part) was $1,000 ÷ 5 = $200, and her total after-tax salary was 6 · $200 = $1,200. But her after-tax salary was 4/5 of her total salary (in blue), so each fifth of her total salary was $1,200 ÷ 4 = $300. So, her total salary was 5 · $300 = <u>$1,500</u>.

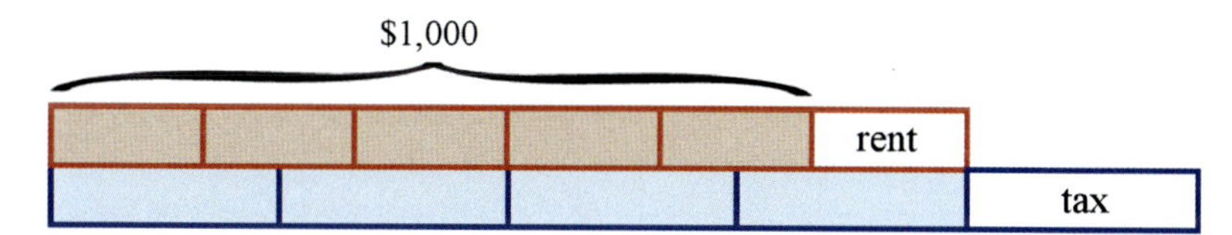

8. a. $45.7 \div 0.02 = 2{,}285$
   b. $928 \div 0.003 \approx 309{,}333.33$
   c. $\frac{5}{8} = 0.625$

Percentage Review, p. 39

1.

a. <u>68%</u> = $\frac{68}{100}$ = <u>0.68</u>	b. 7% = $\frac{7}{100}$ = <u>0.07</u>	c. 15% = $\frac{15}{100}$ = 0.15
d. 120% = $\frac{120}{100}$ = <u>1.20</u>	e. <u>224%</u> = $\frac{224}{100}$ = <u>2.24</u>	f. 6% = $\frac{6}{100}$ = 0.06

2.

percentage / number	**6,100**	**90**	**57**	**6**
1% of the number	61	0.9	0.57	0.06
4% of the number	244	3.6	2.28	0.24
10% of the number	610	9	5.7	0.6
30% of the number	1,830	27	17.1	1.8

3. There are 15 + 5 = 20 skaters. So 15/20 = <u>75%</u> of the skaters are girls

4. a. 75% b. 8% c. 163%

5. Emma's height is 133% of Madison's height. 64/48 = 133

6. The other chair costs $35. 1.4 × 25 = 35

7. 25 is 1/5 of the marbles. There are 5 × 25 = 125 marbles in total, and 4 × 25 = 100 white marbles.

8. 2,000/540 = 0.27 Andrew pays 27% of his salary in taxes.

9. Since 0.80 × $18 = $14.40 and 0.90 × $16 = $14.40, they are both the same price.

10. The area of the square with 2-cm sides is 4 sq. cm. and the area of the square with 4-cm sides is 16 sq cm. Because 4/16 = 0.25, the area of the smaller square is 25% of the area of the larger square.

Percentage Test, p. 41

1.

a. <u>45%</u> = $\frac{45}{100}$ = <u>0.45</u>	b. 179% = $\frac{179}{100}$ = <u>1.79</u>	c. <u>2%</u> = $\frac{2}{100}$ = 0.02

2.

percentage / number	**5,200**	**80**	**9**
1% of the number	52	0.8	0.09
3% of the number	156	2.4	0.27
70% of the number	3,640	56	6.3

3. 57.1%

4. $8.40. First find 30% of $12. It is $3.60. Then subtract that from $12.00.

5. $8.40. First find 20% of $7. Since 10% of $7 is $0.70, then 20% of $7 is $1.40 Add that to the original price of $7.00.

6. Seventy-two T-shirts are not white. You can first calculate 10% of 120, which is 12. Then, 60% of 120 is six times as much, or 72.

7. Twenty percent of the caps are red. There are 2 red caps and 10 in total. Two caps is 2/10 or 20% of the caps.

8. Sixty-seven percent are boys. There are 16 boys, so in total, 16/24 = 2/3 = 67% of the members are boys.

Percentage Test, cont.

9. 144 cm/160 cm = 9/10 = 90%. Annie's height is 90% of Jessie's height.

10. The \$35 jeans discounted by 10% are cheaper. They are \$0.50 cheaper.
    To find the price of \$35 jeans discounted by 10%: First find 10% of \$35. It is \$3.50. Then subtract \$35 − \$3.50 = \$31.50.
    To find the price of \$40 jeans discounted by 20%: First find 20% of \$40. It is \$8. Then subtract \$40 − \$8 = \$32.

11. His salary is \$2,000. If \$400 is 20%, then \$200 is 10%, and \$2,000 is 100% or Andrew's total salary.

12. The total population is 14,000. If 15% is 2,100, divide those by 3 to get that 5% is 700. Then, multiply that 700 by 20 to get 100% or the total population.

Mixed Review 7, p. 43

1. a. 13,054 R23; $13{,}054 \times 26 + 23 = 339{,}427$
   b. 45 R69; $45 \times 145 + 69 = 6{,}594$

2. a. $<$ b. $>$ c. $<$ d. $<$ e. $<$ f. $<$

3. a. $150 - 63 = 87$

   b. $\frac{8}{5} = 1\frac{3}{5} = 1.6$

4. $A = (3x \cdot 5x) + (2x \cdot 2x) = 15x + 4x = 19x^2$; $P = 5x + 3x + 5x + 1x + 2x + 2x + 2x = 20x$

5. a. $4y + 7$
   b. $8r^3$

6. $30 \div 2.54 \approx 11.8$. Technically, the ruler is would be 11.8 inches long, but in reality, 30-cm rulers are made long enough to show 12 inches.

7. The expressions that have the value of 6 are: a, c, d, f, g, i, j, and l.

8. a. The box weighs 280 grams.
   b. They would need to buy four boxes of paper clips.

9. $\frac{\$2}{5 \text{ min}} = \frac{\$6}{15 \text{ min}} = \frac{\$8}{20 \text{ min}} = \frac{\$10}{25 \text{ min}} = \frac{\$24}{1 \text{ hr}}$

10. The sides of the rectangle make some multiple of the ratio 1:7, so let's call them x and $7x$. The perimeter is thus $x + 7x + x + 7x = 16x$. If the perimeter measures 120mm, then $x = 120\text{mm}/16 = 7.5$. So the width is $x = 7.5$mm, and the height is $7x = 52.5$mm.

11. Since 8:10 = 4:5 = 20:25, Gary can expect to make 20 baskets for every 25 shots.

12.

a.	b.	c.
$312 = x + 78$ $312 - 78 = x$ $234 = x$	$\frac{z}{2} = 60 + 80$ $\frac{z}{2} = 140$ $z = 2 \cdot 140$ $z = 280$	$7y - 2y = 45$ $5y = 45$ $y = 45 \div 5$ $y = 9$

13. $m = 0.3048 \cdot 89 \text{ ft} = 27.1272 \text{ m} \approx 27.13 \text{ m}$

14.

Distance	8	24 miles	40	48	120	144
Time	10 min	30 min	50 min	1 hour	2 1/2 hours	3 hours

Mixed Review 8, p. 45

1. a. 0.00392 b. 5.0015
   c. 0.000023 d. 12.012

2. a. 16/1,000,000 b. 29381/10,000 c. 39,402/100,000

3.

a. $1 - 0.05 = 0.95$	b. $0.1 - 0.05 = 0.05$	c. $1.1 - 0.05 = 1.05$

4.

	2.97167	0.046394	2.33999	1.199593
the nearest tenth	3.0	0.0	2.3	1.2
the nearest thousandth	2.972	0.046	2.340	1.200

5. Final answers do not vary, but the ways to get there can vary.

a. $\frac{5.6}{0.4} = \frac{56}{4} = 14$	b. $\frac{4}{0.02} = \frac{40}{0.2} = \frac{400}{2} = 200$	c. $\frac{0.9}{0.003} = \frac{9}{0.03} = \frac{90}{0.3} = \frac{900}{3} = 300$

6. $\frac{320 \text{ people}}{1{,}200 \text{ people}} = \frac{4}{15} = \frac{40 \text{ people}}{150 \text{ people}}$

   a. The ratio of people who like mashed potatoes best to the total number of people interviewed is 4:15.
   b. Of a group of 150 people, we would expect about 40 to prefer mashed potatoes.

7.

a. $\frac{14 \text{ km}}{20 \text{ min}} = \frac{3.5 \text{ km}}{5 \text{ min}} = \frac{31.5 \text{ km}}{45 \text{ min}}$	b. $\frac{\$33.60}{8 \text{ bottles}} = \frac{\$4.20}{1 \text{ bottle}} = \frac{\$42}{10 \text{ bottles}}$

8. $\frac{2 \text{ kg}}{120 \text{ m}^2} = \frac{1 \text{ kg}}{60 \text{ m}^2} = \frac{5 \text{ kg}}{300 \text{ m}^2}$ The lawn has an area of $15 \text{ m} \cdot 20 \text{ m} = 300 \text{ m}^2$, so you would need 5 kg of fertilizer.

9.

x	2	3	4	5	6	7
y	0	2	4	6	8	10

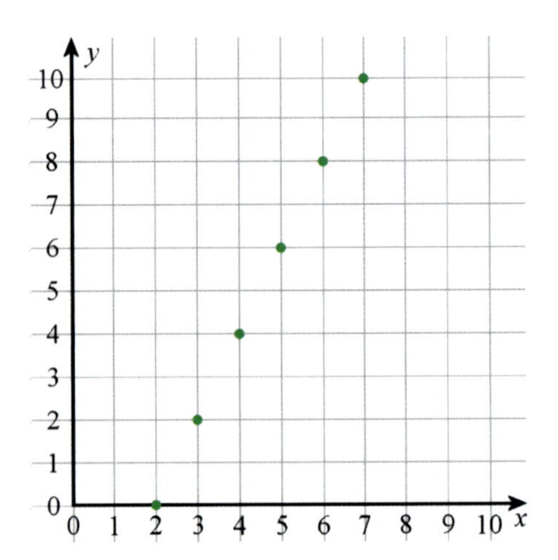

10. If two-thirds of a stick is 50 cm long, then each third is 25 cm long, and the whole stick is three-thirds long, or $3 \cdot 25$ cm = 75 cm.

11. Two gallons is 256 ounces. Then, 256 oz ÷ 6 oz = 42 servings with 4 oz left over.

12. a. Hannah 151 cm; Erica 136 cm
    b. Hannah 175 cm or 1,750 mm; Erica 160 cm or 1,600 mm.
    c. Hannah 165 cm or 1.65 m; Erica 150 cm or 1.5 m.

13. Every afternoon Erica bicycles 5 miles (8.0 km) to the horse ranch. Erica takes care of a horse that is 15 *hands*, or 60 inches (1.5 m), tall. She likes to go riding on a trail that is 4 mi 500 ft (6.6 km) long.

(5 mi · 1.6093 km/mi = 8.0465 km)
(60 in · 2.54 cm/in = 152.4 cm)
(21,620 ft · 0.3048 m/ft ≈ 6,590 m)

Prime Factorization, GCF, and LCM Review, p. 47

1.

a. $3 \times 3 \times 3 \times 3$	b. 2×13	c. 5×13
d. $3 \times 2 \times 2 \times 2 \times 2 \times 2$	e. $2 \times 2 \times 31$	f. $2 \times 3 \times 3 \times 5 \times 5$

2.

a. $\frac{28}{84} = \frac{\overset{1}{\cancel{4}} \times \overset{1}{\cancel{7}}}{\underset{3}{\cancel{21}} \times \underset{1}{\cancel{4}}} = \frac{1}{3}$	b. $\frac{75}{160} = \frac{\overset{1}{\cancel{5}} \times 15}{\underset{2}{\cancel{10}} \times 16} = \frac{15}{32}$
c. $\frac{222}{36} = \frac{\overset{1}{\cancel{6}} \times 37}{6 \times \underset{1}{\cancel{6}}} = \frac{37}{6} = 6\frac{1}{6}$	d. $\frac{48}{120} = \frac{\overset{1}{\cancel{6}} \times \overset{4}{\cancel{8}}}{\underset{2}{\cancel{12}} \times \underset{5}{\cancel{10}}} = \frac{4}{10} = \frac{2}{5}$

3. a. 21 b. 40 c. 66 d. 24

4. a. 8 b. 25 c. 16 d. 6

5. a. 25, 50, 75, 100, 125, and 150 are multiples of 25.
   1, 2, 5, 10, 25, and 50 are factors of 50.
   Each number has an infinite number of multiples.
   Each number has a greatest factor.
   If the number x divides into another number y, we say x is a factor of y.

   b. Answers will vary. Please check the students' work.
   Any five of these will work: 75, 90, 105, 120, 135, 150, 165, 180

   c. Answers will vary. Please check the students' work.
   Example: 28, 56, 112, 224, 448. The LCM of 4 and 7 is 28.

6.

a. GCF of 12 and 21 is 3. $12 + 21 = 3 \cdot 4 + 3 \cdot 7 = 3(4 + 7)$
b. GCF of 45 and 70 is 5. $45 + 70 = 5(9 + 14)$

7.

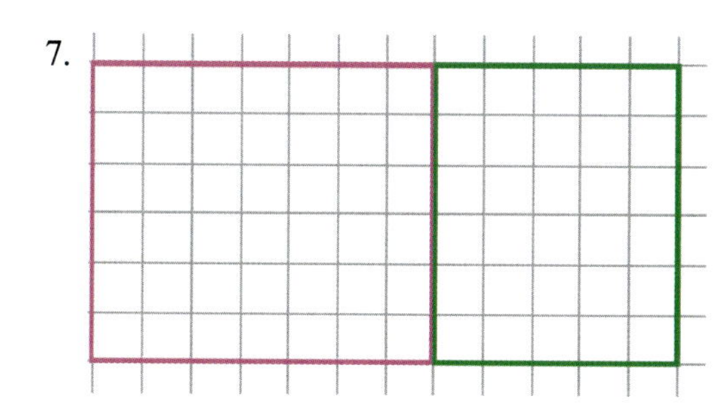

Prime Factorization, GCF, and LCM Test, p. 49

1. a. $2 \cdot 2 \cdot 2 \cdot 7$ b. $2 \cdot 3 \cdot 3 \cdot 5$ c. 101 is a prime number

2. a. 24 b. 12

3. a. 2 b. 7

4. Answers vary. For example 30, 60, 90, and 120.

5. $40 \times 2 = 80$

6. 1 is a factor of all numbers.

7. They are prime numbers so their greatest common factor is 1.

8.

a. GCF of 24 and 30 is 6. $24 + 30 = 6 \cdot 4 + 6 \cdot 5 = 6(4 + 5)$
b. GCF of 22 and 121 is 11. $22 + 121 = 11(2 + 11)$

9.

a. $\frac{124}{72} = \frac{31}{18} = 1\frac{13}{18}$	b. $\frac{65}{105} = \frac{13}{21}$

Mixed Review 9, p. 51

1. a. 30,000 b. 343 c. 1,250

2. a. $109{,}200 = 1 \cdot 10^5 + 9 \cdot 10^3 + 2 \cdot 10^2$
   b. $7{,}002{,}050 = 7 \cdot 10^6 + 2 \cdot 10^3 + 5 \cdot 10^1$

3. The pieces are 3 ft 4.5 in (or 40.5 in.) and 5 ft 7.5 in. (or 67.5 inches) long.
   In inches, the board is 108 inches. The ratio of 3:5 means we think of the board as being in 8 "parts". 108 in. ÷ 8 = 13.5 in.
   So, in a ratio of 3 to 5, the first piece is 3 × 13.5 in. = 40.5 in., and the second piece is 5 × 13.5 in. = 67.5 in.

4.

a. $\frac{16}{0.4} = \frac{160}{4} = 40$	b. $\frac{7}{0.007} = \frac{7{,}000}{7} = 1{,}000$	c. $\frac{99}{0.11} = \frac{9{,}900}{11} = 900$

5.

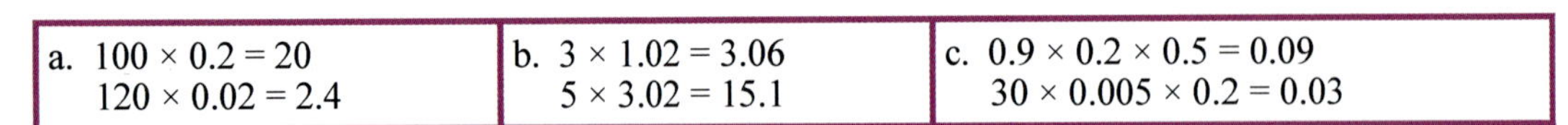

a. $100 \times 0.2 = 20$ $120 \times 0.02 = 2.4$	b. $3 \times 1.02 = 3.06$ $5 \times 3.02 = 15.1$	c. $0.9 \times 0.2 \times 0.5 = 0.09$ $30 \times 0.005 \times 0.2 = 0.03$

6. a.

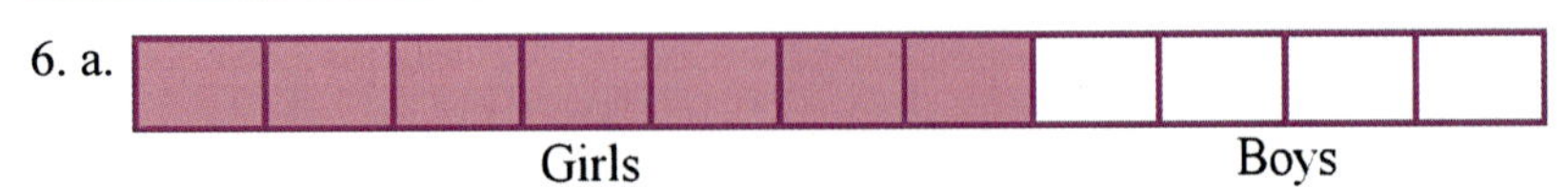

   b. The ratio of boys to all students is 4:11.
   c. There are 748 ÷ 11 × 7 = 476 girls.
      There are 748 ÷ 11 × 4 = 272 boys.

7. Liz's height is 83.3% of her dad's height.

8. Their total budget for the month was $2,250.

9. $180 × 0.80 = $144
   $155 × 0.90 = $139.50, which is cheaper.

Mixed Review 9, cont.

10.

a. $2(7m + 4) = 2 \cdot 7m + 2 \cdot 4 = 14m + 8$	b. $10(x + 6 + 2y) = 10 \cdot x + 10 \cdot 6 + 10 \cdot 2y = 10x + 60 + 20y$

11. a. $\frac{5s + 8}{7}$ b. $(n + 11)^3$

c. $8 + y$ d. $\frac{x}{y^2}$

12. The numbers −2, −1, 0, 1, and 2 fulfill the inequality (any number less than 3 from the given set).

13. a. 3.093 b. 0.206

Mixed Review 10, p. 53

1. a. 80% b. 85% c. 45%

2. a. 0.8 b. 0.85 c. 0.45

3. a.

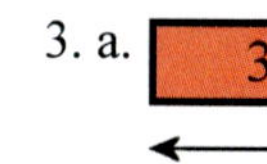

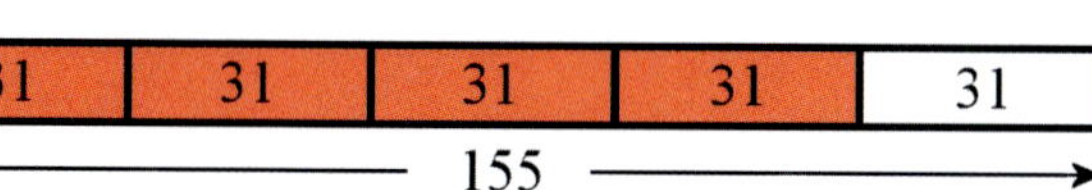

b. 4/5 were basic calculators.
c. 80% were basic calculators.
d. There were 31 scientific calculators.

4.

Distance	2 km	4 km	5 km	16 km	20 km	24 km	40 km	60 km	70 km
Time	6 min	12 min	15 min	48 min	1 hour	1 h 12 min	2 hours	3 hours	3 1/2 hours

5. $53.75
The price of the other flash drive is $25 + $2.50 + $1.25 = $28.75. Buying both, the cost is $25 + $28.75 = $53.75.

6.

a. $32t + 8 = 8(4t + 1)$	b. $8 + 12x = 4(2 + 3x)$
c. $15y + 45 = 3(5y + 15)$	d. $35 + 42w = 7(5 + 6w)$

7. Her score is 35/40 = 7/8 = 87.5%.

8. He has 9 cars left. One-fifth of his cars is 18 cars, so that is how many Jack kept. Then he gave half of those to his brother.

9.

a. $10{,}000 \times 0.092 = 920$	b. $1{,}000 \times 0.0004 = 0.4$
c. $456.29 \div 1{,}000 = 0.45629$	d. $63 \div 10^5 = 0.00063$

10. a. 15.34 m b. 0.334 L c. 900 g

11. a. You need to first choose a variable for the unknown. Let b be the cost of the camera bag.
The equation is b + $85 = $162 or $162 − $85 = b. The camera bag costs $77.
b. Let p be the price of one towel. $8t$ = $52. One towel cost $6.50.

12. Three shirts cost $32.90. $14.10 ÷ 3 × 7 = $32.90

13.

a. 79 oz = 4 lb 15 oz b. 4 ft 11 in = 59 in	c. 7.82 qt = 1.96 gal d. 0.265 mi = 466.4 yd	e. 2.54 lb = 40.64 oz f. 6.8 ft = 6 ft 9.6 in

Fractions Review, p. 55

1.

a. $\frac{5}{12} + \frac{4}{12} = \frac{9}{12} = \frac{3}{4}$	b. $\frac{30}{42} + \frac{7}{42} = \frac{37}{42}$	c. $\frac{64}{40} + \frac{35}{40} = \frac{99}{40} = 2\frac{19}{40}$

2.

a. $6\frac{2}{3} \rightarrow 6\frac{4}{6}$ $-\ 2\frac{1}{6} \rightarrow -\ 2\frac{1}{6}$ $4\frac{1}{2}$	b. $7\frac{1}{6} \rightarrow 6\frac{35}{30}$ $-\ 2\frac{3}{5} \rightarrow -\ 2\frac{18}{30}$ $4\frac{17}{30}$	c. $8\frac{9}{11} \rightarrow 8\frac{27}{33}$ $-\ 4\frac{1}{3} \rightarrow -\ 4\frac{11}{33}$ $4\frac{16}{33}$

3.

a. $\frac{3}{4} \times \frac{2}{3} = \frac{6}{12} = \frac{1}{2}$	b. $\frac{1}{5} \times \frac{1}{2} = \frac{1}{10}$	c. $\frac{2}{3} \times \frac{3}{5} = \frac{6}{15} = \frac{2}{5}$

4.

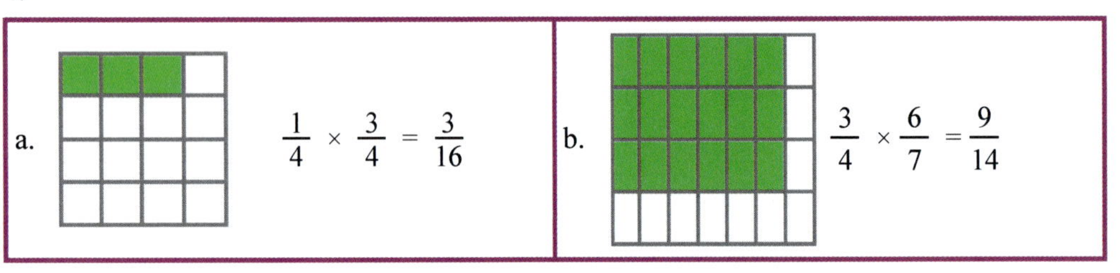

a. $\frac{1}{4} \times \frac{3}{4} = \frac{3}{16}$

b. $\frac{3}{4} \times \frac{6}{7} = \frac{9}{14}$

5. a. $\frac{3}{2} \times \frac{1}{5} = \frac{3}{10}$ b. $\frac{1}{5} \times \frac{1}{7} = \frac{1}{35}$ c. $\frac{1}{4} \times \frac{1}{3} = \frac{1}{12}$

6.

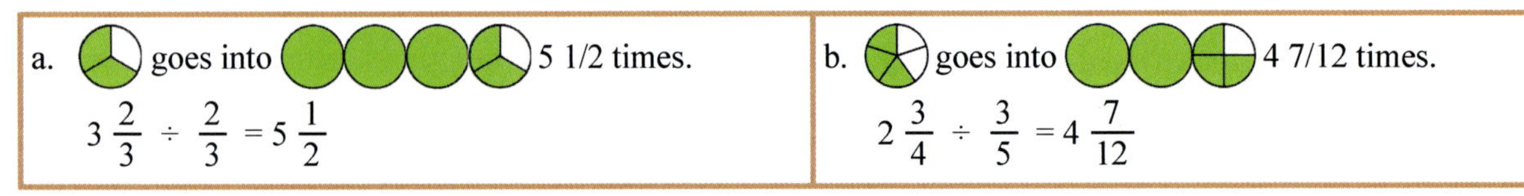

a. $3\frac{2}{3} \div \frac{2}{3} = 5\frac{1}{2}$

b. $2\frac{3}{4} \div \frac{3}{5} = 4\frac{7}{12}$

7. Dividing a number by 4 is the same as multiplying it by 1/4.
   Example: 16 ÷ 4 = 4 or 16 × (1/4) = 4.

8. a. 3 1/3 b. 1 3/7 c. 3 3/5

9. a. The area of the garden is 32 13/16 square feet.
   b. The perimeter of the garden is 23 3/4 feet.

10. Answers will vary. Please check the students' work.
    Example: Three people shared 9/12 of a pie. How much pie did each person get? $\frac{9}{12} \div 3 = \frac{3}{12}$

11. You can cut 28 pieces with a little bit of string left over.
    You can divide to find the answer, but since the two quantities are in different units, one of them needs converted so that both are in feet or that both are in inches. For example, we can change 10 ft into 120 inches, and then divide: 120 in ÷ (4 1/4 in) = 120 in ÷ (17/4 in) = 120 × (4/17) = 480/17 = 28 4/17.

12. One piece was 1 7/8 inches long and the other piece was 13 1/8 inches long.
    The ratio of 1:7 means there are 8 equal parts. Divide 15 inches by 8 to find the length of one part: 15 in ÷ 8 = 15/8 in = 1 7/8 in.

13. The wingspan is 40 feet 3 3/4 inches. Multiply 15 × (32 1/4 in) = 15 × 32 in + 15/4 in = 483 3/4 in = 40 ft 3 3/4 in.

Fractions Review, cont.

14. There were 96 people in the class. One-sixth of the students was 16 students, so all the students are six times that.

15. The number is 400. Since two-fifths of the number is 160, one-fifth of it is 80. Multiply that by 5 to get the number.

16. The total kiwi harvest was 82 1/2 pounds.
The amount given to the son was 11 lb. Two-fifths of the harvest was 33 lb. Therefore, one-fifth of the harvest was 16.5 lb, and the total harvest was five times that, or 82.5 lb.

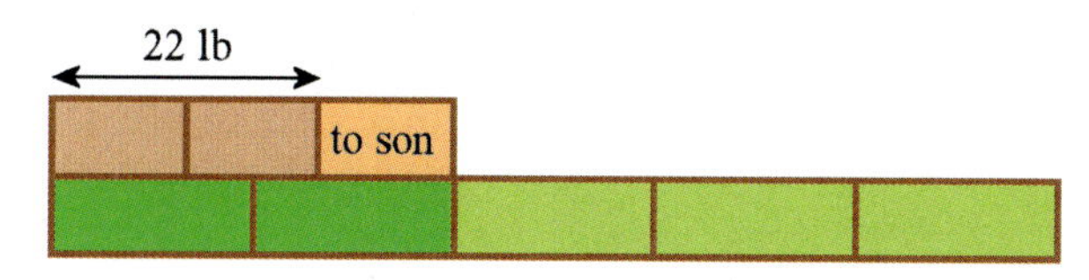

Puzzle corner:
a. 1/240 b. 3/4

Fractions Test, p. 58

1.

a. $\frac{5}{12} + \frac{6}{12} + \frac{10}{12} = \frac{21}{12} = \frac{7}{4} = 1\frac{3}{4}$	b. $\frac{35}{63} - \frac{18}{63} = \frac{17}{63}$
c. $2\frac{18}{60} + 2\frac{55}{60} = 4\frac{73}{60} = 5\frac{13}{60}$	d. $7\frac{3}{15} - 5\frac{7}{15} = 1\frac{11}{15}$

2. 1/4 × 3/4 = 3/16 of the original pizza left.
Since Joe ate 3/4 of what had been left, Joe did not eat 1/4 of what had been left. So, what remains is 1/4 of the 3/4 of the pizza. Or, you can calculate that the part Joe ate was 3/4 × 3/4 = 9/16 of the original pizza, and the family had eaten 1/4 = 4/16 of the original pizza, which means a total of 13/16 of the pizza had been consumed, and there was 3/16 of it left.

3. You will get twenty-one 1/4-kg servings of meat with 1/12 kg or 83 grams left over.
5 1/3 ÷ (1/4) = (16/3) ÷ (1/4) = 16/3 × 4 = 64/3 = 21 1/3. So, you get 21 servings, and 1/3 of a serving.
Since one serving is 1/4 kg, 1/3 of a serving is 1/12 kg.

4. Each piece is 1 4/9 ft long. 4 1/3 ÷ 3 = 13/3 × (1/3) = 13/9 = 1 4/9 ft.

5.

a. 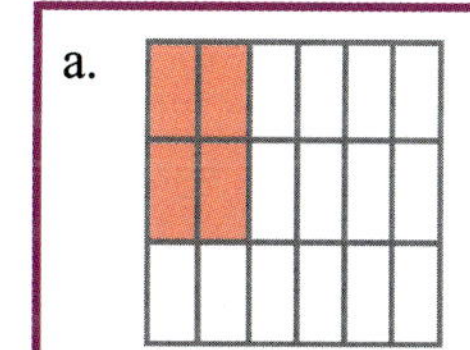 $\frac{2}{6} \times \frac{2}{3} = \frac{4}{18} = \frac{2}{9}$	b. 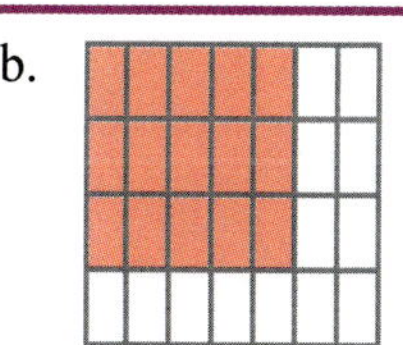 $\frac{3}{4} \times \frac{5}{7} = \frac{15}{28}$

6.

a. $\frac{6}{7} \times \frac{5}{1} = \frac{30}{7} = 4\frac{2}{7}$	b. $\frac{12}{13} \times \frac{3}{7} = \frac{36}{91}$

7. The carpet costs $269.50. You can calculate the area of the room using either fractions or decimals. Using fractions, the area of the room is 11 ft × 8 3/4 ft = (11 ft × 8 ft) + (11 ft × 3/4 ft) = 88 sq. ft. + 33/4 sq. ft = 88 sq. ft + 8 1/4 sq. ft = 96 1/4 sq. ft. Using decimals, the area of the room is 11 ft × 8.75 ft = 96.25 sq. ft.
To calculate the price, use decimal multiplication: 96.25 × $2.80 = $269.50

8. Answers vary. Please check the students' answers. For example, there are 2 1/2 pizzas left, and three people share them evenly. How much pizza does each get? $2\frac{1}{2} \div 3 = \frac{5}{2} \times \frac{1}{3} = \frac{5}{6}$. Each person gets 5/6 of a pizza.

9. You can get six pieces with 1 1/2 ft left over.
12 ft ÷ (1 3/4 ft) = 12 ÷ (7/4) = 12 × (4/7) = 48/7 = 6 6/7. This means you get 6 pieces and 6/7 of a piece. The six pieces are a total of 6 × (1 3/4 ft) = 6 18/4 ft = 10 1/2 ft, so since the string was 12 ft, you will have 1 1/2 ft of string left over.

10. Aiden now has $50.40.
The ratio of 2:3 means Aiden got 3/5 of the reward. So, he got $120 ÷ 5 × 3 = $72. Then, since Aiden gave 3/10 of his money to his dad, he was left with 7/10 of it. $120 ÷ 10 × 7 = $50.40 (or you can also calculate it as $72 × 0.7).

Mixed Review 11, p. 60

1. a. There are 12 inches in one foot, so in one mile we have about 5,000 × 12 = 60,000 inches. Or, you can estimate it as 5,300 × 12 = 63,600 inches
   b. There are exactly 63,360 inches in one mile.

2. There are 16 ounces of juice concentrate and 48 ounces of water. There are a total of 8 "parts". Each part is 64 oz ÷ 8 = 8 oz. So, there are 2 × 8 oz = 16 oz of juice concentrate and 6 × 8 oz = 48 oz of water.

3.

a. $\frac{\$80}{4\text{ hr}} = \frac{\$20}{1\text{ hr}} = \frac{\$60}{3\text{ hr}} = \frac{\$5}{15\text{ min}}$	b. $\frac{2\text{ m}^2}{5\text{ min}} = \frac{10\text{ m}^2}{25\text{ min}} = \frac{120\text{ m}^2}{5\text{ hours}} = \frac{250\text{ m}^2}{10\text{ hr }25\text{ min}}$

4. There are 24 grams of salt and 1,176 grams of water.
   The mixture weighs 1,200 g. So, 1% of it is 12 g and 2% of it is 24 g.

5. The train left at 12:50 p.m.
   The table below helps you find how much time it took the train to travel 165 miles:

3 miles	15 miles	45 miles	90 miles	135 miles	150 miles	165 miles	180 miles
2 min	10 min	30 min	1 h	1 1/2 h	1 h 40 min	1 h 50 min	2 h

   Going 1 h 50 min backwards from 2:40 p.m. gives us 12:50 p.m.

6.

a.	b.	c.
10 × 0.3909 = 3.909 1,000 × 4.507 = 4507	1.08 × 100 = 108 0.0034 × 10^4 = 34	10^6 × 8.02 = 8020000 10^5 × 0.004726 = 472.6
d. 0.93 ÷ 100 = 0.0093 48 ÷ 10 = 4.8	e. 3.04 ÷ 1,000 = 0.00304 450 ÷ 10^4 = 0.045	f. 98.203 ÷ 10^5 = 0.00098203 493.2 ÷ 10^6 = 0.0004932

7. a. 65 = 5 × 13   b. 75 = 5 × 5 × 3   c. 82 = 2 × 41

8. 651

9.

a. GCF of 16 and 42 is 2 16 + 42 = 2(8 + 21)	b. GCF of 98 and 35 is 7 98 + 35 = 7 (14 + 5)

10.

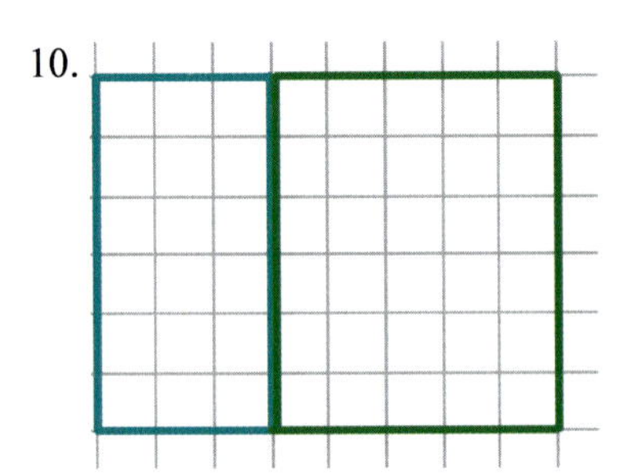

Mixed Review 11, cont.

11.

x	3	4	5	6	7	8
y	1	3	5	7	9	11

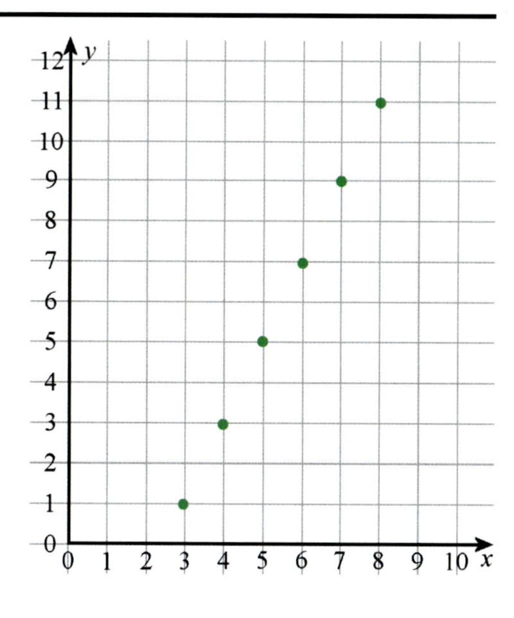

Mixed Review 12, p. 62

1. a. 8 b. 18

2. a. 1 b. 12

3.

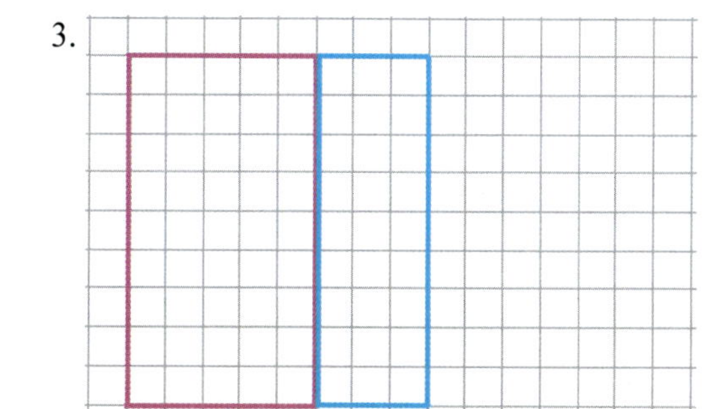

4. Each person gets 7.11 ounces of ice cream. Two quarts is 64 ounces, so each person gets $2 \times 32 \div 9 = 7.11$ oz.

5. a. 11^2 gives us the <u>area</u> of a <u>square</u> with a side length of 11 units.
   b. 3×5^2 gives us the <u>area</u> of <u>3</u> <u>squares</u> with a side length of <u>5</u> units.
   c. 4×0.4^3 gives us the <u>volume</u> of <u>4</u> <u>cubes</u> with an edge length of <u>0.4</u> units.

6. a. 26,020,000 b. 1,000,200,807

7. a. $3 \times 3 \times 11$ b. $2 \times 2 \times 2 \times 2 \times 7$ c. $2 \times 2 \times 2 \times 5 \times 5$

8. The total capacity is 49.5 liters.
   The smaller container holds $8.5 \times 0.60 = 5.1$ liters. The total capacity is then $3 \times 8.5 + 4 \times 5.1 = 45.9$ liters.

9. Samantha earned \$25 more than George. Samantha earned $\$100 \div 8 \times 5 = \$62.50$ and George earned \$37.50. Or, you can think this way: Samantha earned 5 "parts" and George 3 "parts"; therefore the difference in their earnings is 2 "parts". Each part is $\$100 \div 8 = \$12.50$, so two parts is \$25.

10. a. $\frac{6}{7s}$ b. $11 - 2x$

    c. $(x + 2)^2$ d. $(5m)^3$

    e. $\frac{2t^2}{s - 1}$ f. $18 - y$

Mixed Review 12, cont.

11. $2x + 194 = 388$
$2x = 388 - 194$
$x = 97$

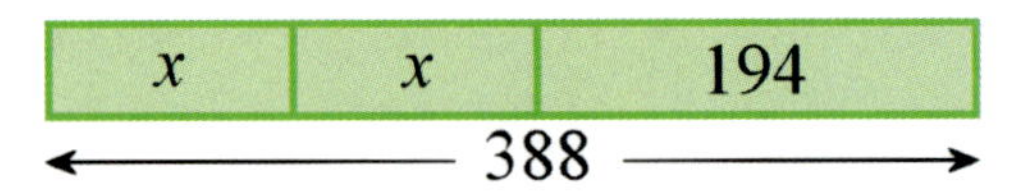

12.

a. $4 \times 0.7 = 2.8$ b. $50 \times 0.003 = 0.15$	c. $3 \times 1.06 = 3.18$ d. $100 \times 0.009 = 0.9$	e. $10^5 \times 0.08 = 8{,}000$ f. $40 \times 0.004 = 0.16$

Integers Review, p. 64

1.

a. $-1 > -7$	b. $2 > -2$	c. $-6 < 0$	d. $8 > -3$	e. $-8 < -3$

2.

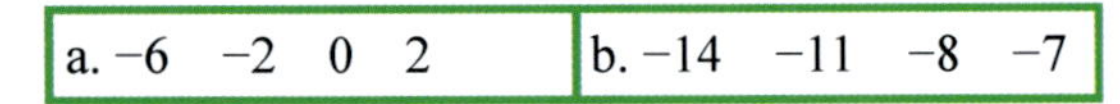

a. −6 −2 0 2	b. −14 −11 −8 −7

3. a. $-\$12 > -\$18$
   b. $-5°C < 2°C$
   c. $16 \text{ m} > -6 \text{ m}$

4. a. 11 b. 2 c. 0 d. 19 e. −7

5. a. $-9 + 6 = -3$ b. $-2 + 5 = 3$

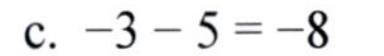
c. $-3 - 5 = -8$

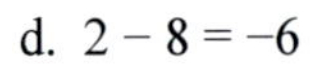
d. $2 - 8 = -6$

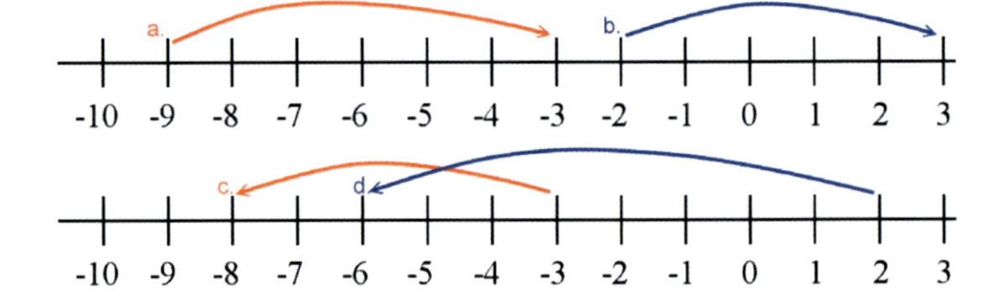

6. a. You are at $^{-}10$. You jump 6 to the right. You end up at $^{-}4$.
   b. You are at $^{-}5$. You jump 8 to the right. You end up at 3.
   c. You are at 3. You jump 7 to the left. You end up at $^{-}4$.
   d. You are at $^{-}11$. You jump 3 to the left. You end up at $^{-}14$.

$^{-}10 + 6 = {^{-}4}$

$^{-}5 + 8 = 3$

$3 - 7 = {^{-}4}$

$^{-}11 - 3 = {^{-}14}$

7.

a. $2 + (-8) = {^{-}6}$ $(-2) + 8 = 6$	b. $-2 + (-9) = {^{-}11}$ $2 - 8 = {^{-}6}$	c. $1 + (-7) = {^{-}6}$ $-4 - 5 = {^{-}9}$	d. $5 - (-2) = 7$ $-3 - (-4) = 1$

8. a. May has \$35. She wants to purchase a guitar for \$85.
   That would make her money situation to be $^{-}$\$50.
   b. A fish was swimming at the depth of 6 ft. Then he sank 2 ft.
   Then he sank 4 ft more. Now he is at the depth of $^{-}$12 ft.
   c. Elijah owed his dad \$20. Then be borrowed another \$10.
   Now his balance is $^{-}$\$30.
   d. The temperature was −13°C and then it rose 5°.
   Now the temperature is −8 °C.

$\$35 - \$85 = {^{-}\$50}$

$^{-}6 \text{ ft} - 2 \text{ ft} - 4 \text{ ft} = {^{-}12} \text{ ft}$

$^{-}\$20 - \$10 = {^{-}\$30}$

$-13°C + 5°C = -8°C$

9. a. $|-17|$ b. $-(-11)$

10. d. balance $< -\$50$

Integers Review, cont.

11.

x	−5	−4	−3	−2	−1	0	1	2
y	9	8	7	6	5	4	3	2

x	3	4	5	6	7	8	9
y	1	0	−1	−2	−3	−4	−5

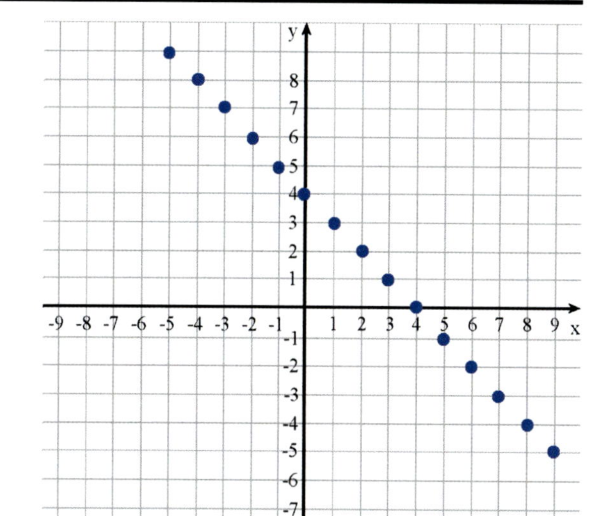

12.

a. $-2 + (-6) = -8$ $3 + (-5) = -2$	b. $4 + (-4) = 0$ $-6 - 6 = -12$	c. $5 - 7 = -2$ $3 + (-2) = 1$

13. (−9, −6), (−6, −6), (−9, −3), and (−3, 0).

14. a. The distance is 12 + 15 = 27 units.
    b. The distance is 21 − 15 = 6 units.

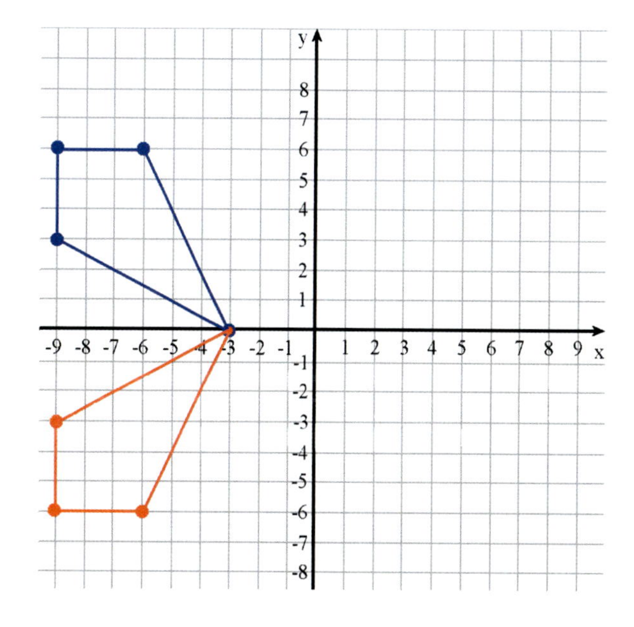

Integers Review, cont.

15. a. The points $(-7, -3)$, $(-1, -7)$, $(-1, -1)$, and $(-4, -6)$

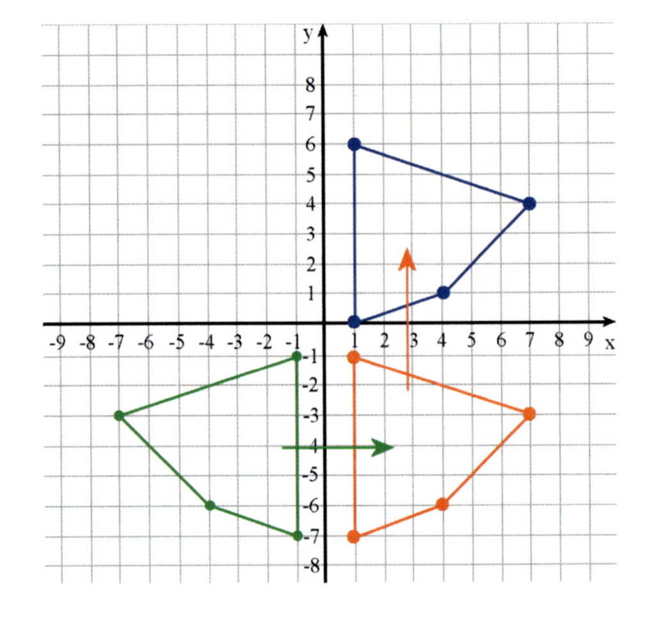

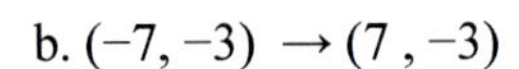

b. $(-7, -3) \rightarrow (7, -3)$

$(-1, -7) \rightarrow (1, -7)$

$(-1, -1) \rightarrow (1, -1)$

$(-4, -6) \rightarrow (4, -6)$

c. $(7, 4)$ $(1, 0)$ $(1, 6)$ $(4, 1)$

Integers Test, p. 67

1. $-5, -3, 0, 3$

2. a. $-7 + 2 = -5$

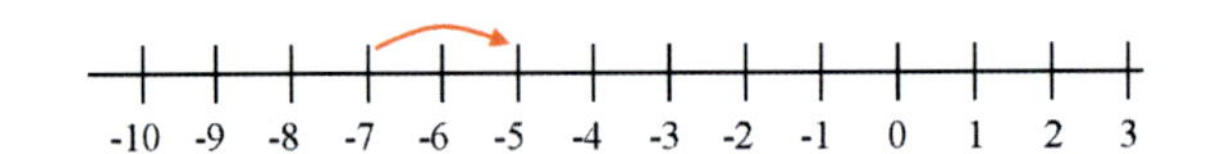

b. $-3 + 6 = 3$

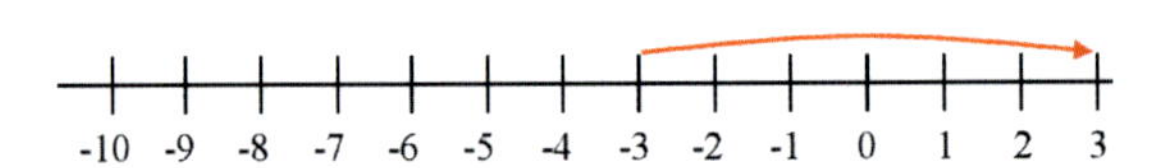

c. $-1 - 5 = -6$

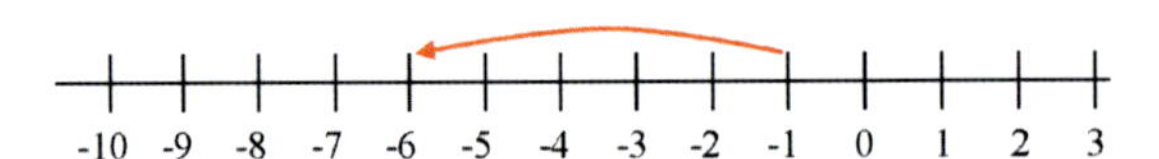

d. $2 - 7 = -5$

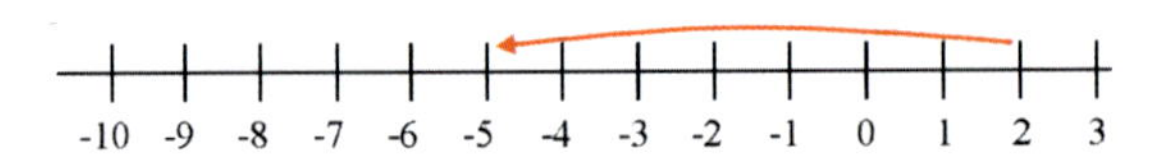

3.

a.	b.	c.	d.
$3 + (-7) = -4$	$(-1) + (-9) = -10$	$4 + (-5) = -1$	$8 - (-2) = 10$
$(-3) + 7 = 4$	$1 - 9 = -8$	$-4 - 5 = -9$	$-8 - (-2) = -6$

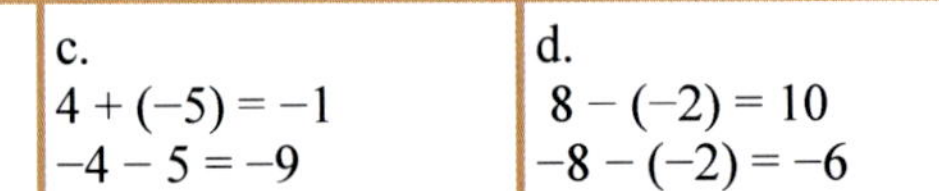

4. a. $|-9|$
   b. -43
   c. Henry's balance $> -\$20$. Students may also include that Henry's balance < 0, but I would not require it in this test. (By the way, usually, these two statements are written as $-\$20 <$ Henry's balance < 0.)
   d. temperature < -10

5. a. Now her money situation is −\$11. | −\$3 − 8 = −\$11

b. Now the temperature is −3°C. | 1°C − 4°C = −3°C

c. Now it is at the depth of −17 m. | −12 m + 5 m − 10 m = −17 m

6.

x	−7	−6	−5	−4	−3	−2	−1	0
y	−8	−7	−6	−5	−4	−3	−2	−1

x	1	2	3	4	5	6	7	8
y	0	1	2	3	4	5	6	7

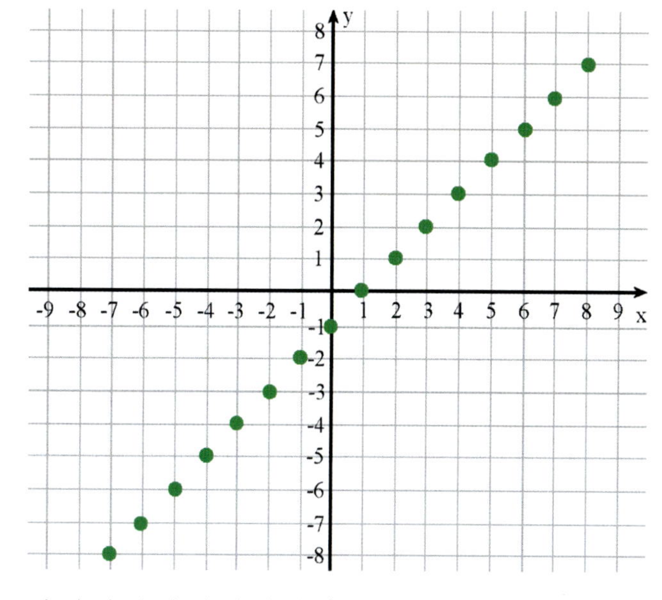

7. a. See the grid on the right.
b. See the grid on the right.
c. The new vertices are (−2, 6), (0, 2), and (4 , 4).

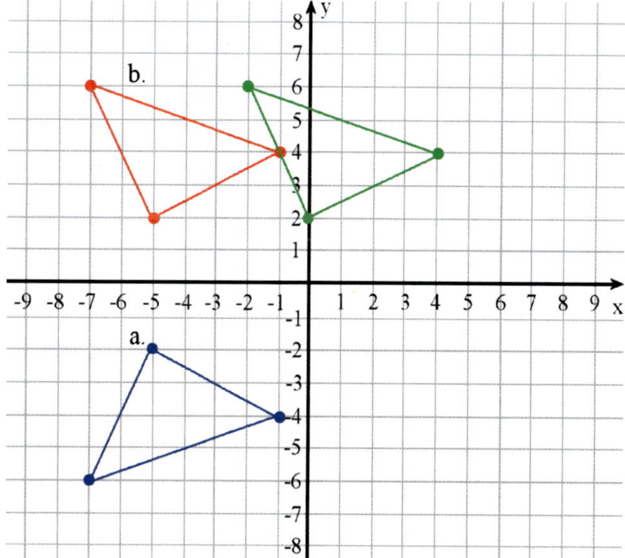

Mixed Review 13, p. 69

1. a. The unknown is Shelly's age so choose a variable for it. Let s be Shelly's age. We get the equation $54 - 12 = s$. Solution: $s = 42$
   b. The unknown is the number of tulips so choose a variable for it, such as n. The equation is $n \times \$2.15 = \$45.15$ or you can write this as $2.15n = 45.15$. Solution: $n = \$45.15 \div 2.15 = 21$. Bob bought 21 tulips for his wife.

2. $512 = 2^9$

3.

a. $\frac{5}{5} \div \frac{4}{5} = 1\frac{1}{4}$	b. $\frac{8}{8} \div \frac{3}{8} = 2\frac{2}{3}$
c. $\frac{3}{4} \times \frac{2}{3} = \frac{1}{2}$	d. $\frac{2}{9} \times \frac{3}{4} = \frac{1}{6}$

4. a. Fifty packages of dominoes weigh 9 lb 6 oz. Multiply to find the total weight: 50×3 oz = 150 oz = 9 lb 6 oz.
   b. The quality dominoes weigh 14 lb 10 oz more than the box with cheap dominoes.

5.

a. $\frac{23}{8} \div \frac{2}{5} = 7\frac{3}{16}$	b. $\frac{4}{1} \div \frac{11}{6} = 2\frac{2}{11}$
c. $\frac{5}{1} \div \frac{2}{7} = 17\frac{1}{2}$	d. $\frac{101}{10} \div \frac{3}{4} = 13\frac{7}{15}$

6. Annabelle can type 35 words in a minute, so in 15 minutes she can type $15 \times 35 = 525$ words.

7. Mom is 42 years old. From the model on the right, we can see that each block is 6 years. Therefore, mom is 7×6 years = 42 years old.

8. The area is 360 cm^2.
   The aspect ratio of 5:2 means the two sides make up 7 "parts" and the total perimeter makes 14 parts. Divide the perimeter 84 cm by this 14 to find the length of one part: 84 cm ÷ 14 = 6 cm. The sides are then 5×6 cm = 30 cm and 2×6 cm = 12 cm. The area is 30 cm × 12 cm = 360 cm^2.

9. Keith paid approximately 22% of his salary in taxes. His total salary was \$414 + \$1459 = \$1873. The percentage is \$414 / \$1873 = 0.22103577...

10. a. 180 km : 1 hr   b. 1 kg for \$0.70   b. 24 miles per 1 gallon

11.

a. $\frac{\overset{1}{\cancel{5}}}{\underset{9}{\cancel{36}}} \times \frac{\overset{\overset{2}{\cancel{8}}}{\cancel{24}}}{\underset{\underset{3}{\cancel{9}}}{\cancel{45}}} = \frac{2}{27}$	b. $\frac{\overset{\overset{1}{\cancel{2}}}{\cancel{16}}}{\underset{\underset{3}{\cancel{6}}}{\cancel{30}}} \times \frac{\overset{5}{\cancel{25}}}{\underset{3}{\cancel{24}}} = \frac{5}{9}$	c. $\frac{\overset{\overset{1}{\cancel{2}}}{\cancel{14}}}{\underset{5}{\cancel{25}}} \times \frac{\overset{7}{\cancel{35}}}{\underset{\underset{3}{\cancel{6}}}{\cancel{42}}} = \frac{7}{15}$

Mixed Review 14, p. 71

1.

a. $\frac{23}{24}$	b. $\frac{11}{35}$	c. $\frac{1}{79}$	d. $\frac{1}{100}$	e. $\frac{1000}{3}$

2.

a. $\frac{6}{7} \div \frac{1}{7} = 6$	b. $\frac{29}{20} \div \frac{3}{20} = 9\frac{2}{3}$	c. $5 \div \frac{1}{3} = 15$	d. $7 \div 1\frac{2}{5} = 5$

3. She can fit 24 stickers on the cover.
Along the longer side of the notebook, she can fit (8 1/2 in) ÷ (1 1/4 in) = (17/2) ÷ (5/4) = (17/2) × (4/5) = 17 × 2/5 = 34/5 = 6 4/5 stickers. But, obviously she would not want to use parts of a sticker, so that means she can fit 6 stickers that way. Similarly, along the shorter side, we divide to find how many she can fit: (5 1/2 in) ÷ (1 1/4 in) = (11/2) ÷ (5/4) = (11/2) × (4/5) = 11 × 2/5 = 22/5 = 4 2/5 stickers. So, she can fit 6 stickers one way and 4 the other, a total of 24 stickers.

4.

Expression	the terms in it	coefficient(s)	Constants
$2x + 3y$	$2x$ and $3y$	2 and 3	none
$0.9s$	$0.9s$	0.9	none
$2a^4c^5 + 6$	$2a^4c^5$ and 6	2	6
$\frac{1}{6}f$	$\frac{1}{6}f$	$\frac{1}{6}$	none

5.

Serves (people)	6	12	18	24	30
butter	1/4 cup	1/2 cup	3/4 cup	1 cup	1 1/4 cups
sugar	1/2 cup	1 cup	1 1/2 cups	2 cups	2 1/2 cups
eggs	1	2	3	4	5
flour	3/4 cup	1 1/2 cups	2 1/4 cups	3 cups	3 3/4 cups

6. To serve 100 people you need 4 1/6 cups of butter, 8 1/3 cups of sugar, 17 eggs (it is rather hard to divide an egg), and 12 1/2 cups of flour. You can for example multiply by 4 the ingredient list for 24 people, and then take 1/3 of the ingredients for 12 people.

7.

a. $900 - \frac{1}{6} \cdot 72 = 888$	b. $23 + 3^4 = 104$	c. $\frac{100^3}{100^2} = 100$

8. We can see from the model that the difference of 9 years corresponds to 3 blocks, so one block is 3 years.
a. Marie is 12 years old. Tom is 21 years old.
b. The ratio of Marie's age to Tom's age is 4:7

Tom
Marie
9 years

9.

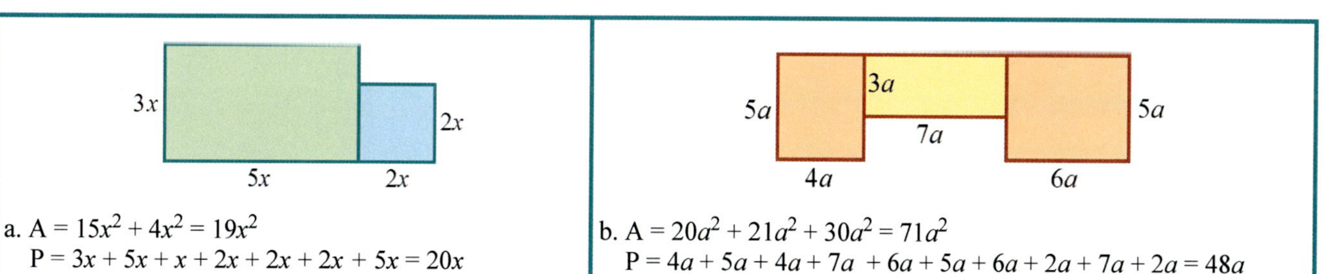

a. $A = 15x^2 + 4x^2 = 19x^2$
$P = 3x + 5x + x + 2x + 2x + 2x + 5x = 20x$

b. $A = 20a^2 + 21a^2 + 30a^2 = 71a^2$
$P = 4a + 5a + 4a + 7a + 6a + 5a + 6a + 2a + 7a + 2a = 48a$

Geometry Review, p. 73

1. The area of the triangle is exactly half of the area of the parallelogram.

2. a. 15 square units
   b. $20 - 2 - 1 - 4 - 3 = 10$ square units

3. a. $6.5 \text{ m} \times 8.5 \text{ m} + 6.5 \text{ m} \times 4.5 \text{ m} \div 2 = 55.25 \text{ m}^2 + 14.625 \text{ m}^2 = 69.875 \text{ m}^2$

   b. The area with green beans is $3.5 \text{ m} \times 3 \text{ m} = 10.5 \text{ m}^2$. It is 10.5 / 69.875 = 15% of the total garden area.

4. a. $13 \text{ cm} \times 8.2 \text{ cm} \div 2 = 53.3 \text{ cm}^2$

   b. $130 \text{ mm} \times 82 \text{ mm} \div 2 = 5330 \text{ mm}^2$ (you can multiply the previous result by 100).

5. a. Check the student's work, as the nets can vary; for example:

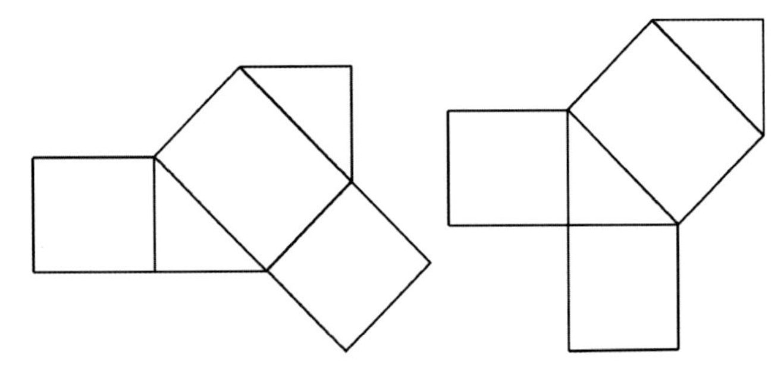

   Surface area: The bottom triangle is $2 \times 19 \text{ in} \times 19 \text{ in} \div 2 = 361$ sq. in.
   The rectangles: $27 \text{ in} \times 21 \text{ in} = 567$ sq. in.; $2 \times 19 \text{ in} \times 21 \text{ in} = 798$ sq. in.
   Total: $361 \text{ in}^2 + 567 \text{ in}^2 + 798 \text{ in}^2 = 1{,}726 \text{ in}^2$

   b.

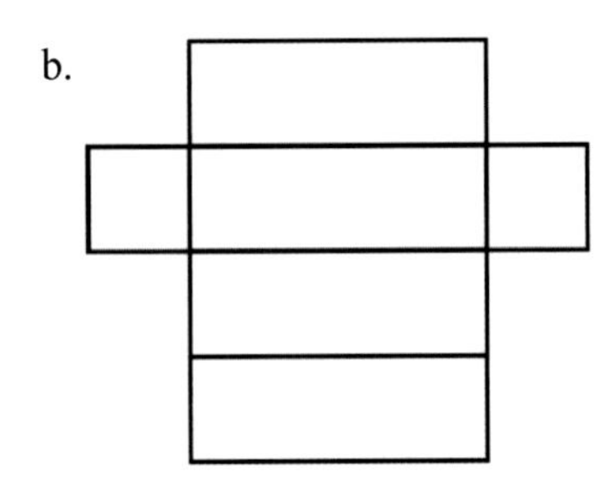

   Surface area: $4 \times 11.75 \text{ in} \times 4 \text{ in} + 2 \times 4 \text{ in} \times 4 \text{ in} = 188$ sq. in. + 32 sq. in. = 220 sq. in.

6. a. triangular prism
   b. rectangular pyramid
   c. tetrahedron (triangular pyramid)

7. triangular prism; surface area: the triangles: $2 \times 3 \text{ in} \times 2\ 5/8 \text{ in} \div 2 = 3 \text{ in} \times 21/8 \text{ in} = 63/8$ sq. in. = 7 7/8 sq. in.
   The lateral sides: $3 \times 3 \text{ in} \times 17 \text{ in} = 153$ sq. in. Total: 160 7/8 sq. in.

8. The figure has 40 cubes, and the volume of each cube is 1/27 m3. The total volume of the figure is $40/27 \text{ m}^3 = 1\ 13/27 \text{ m}^3$. You can also calculate it by multiplying the three dimensions: $2/3 \text{ m} \times 5/3 \text{ m} \times 4/3 \text{ m}$.

9. The volume of one story is $6 \text{ m} \times 12.2 \text{ m} \times 8.5 \text{ m} \div 3 = 207.4 \text{ m}^3$.

10. $50 \text{ cm} \times 30 \text{ cm} \times 40 \text{ cm} \div 5 \times 4 = 48{,}000 \text{ cm}^3 = 48{,}000$ ml = 48 L.

Geometry Test, p. 76

1. Check the student's answers. The results may vary from those given below if the test was not printed at 100%.
   a. The area is 6.4 cm × 3.9 cm = 24.96 cm^2 ≈ 25 cm^2
   b. 64 mm × 39 mm = 2,496 mm^2 ≈ 2,500 mm^2

2. It is 4 × 5 − 2 − 1 − 6 − 1.5 = 9.5 square units

3. a. A trapezoid.
   b. 276 1/4 sq. in. The calculation for the area is (3 in × 13 in) ÷ 2 + (17 in × 13 in) + (5 1/2 in × 13 in) ÷ 2
   = 19 1/2 sq. in. + 221 sq. in. + 35 3/4 sq. in. = 276 1/4 sq. in.

4. The front and back sides are 2 × 2 ft × 1.5 ft = 6 sq. ft. The two other sides are 2 × 1.5 ft × 1.5 ft = 4.5 sq. ft.
   The bottom is 2 ft × 1.5 ft = 3 sq. ft. The total area is 13.5 sq. ft.

5. The volume is 15/32 cubic inches. There are 30 little cubes, each having a volume of 1/64 cubic inches, so the total volume is 30/64 cubic inches = 15/32 cubic inches.

6. The volume is 19 1/2 cubic inches. Calculation: 6 1/2 in × 8 in × 3/8 in = 13/2 in. × 8 in. × 3/8 in. = 52 × 3/8 $in.^3$
   = 13 × 3/2 in^3 = 19 1/2 in^3.

7. a. a square pyramid
   b. The surface area is 66 1/4 in^2.
   Area of the bottom: 5 in. × 5 in. = 25 in^2.
   One of the faces: 5 in. × 4 1/8 in. ÷ 2 = 5 in. × 33/8 in. ÷ 2 = 165/16 in^2 = 10 5/16 in^2.
   Total surface area: 25 in^2 + 4 × 10 5/16 in^2 = 25 in^2 + 41 1/4 in^2 = 66 1/4 in^2.

8. It is a triangular prism. Its net:

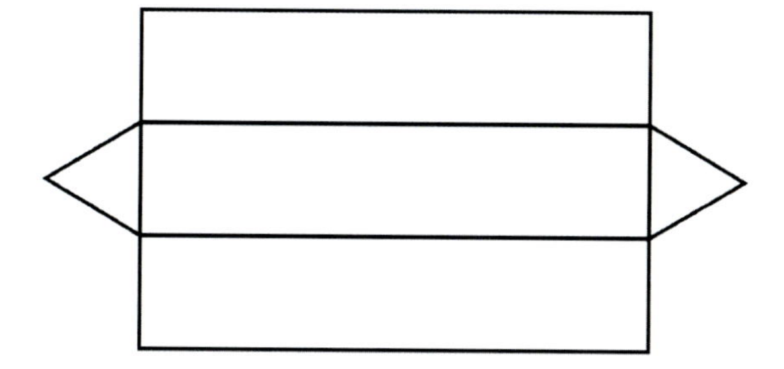

There are other possibilities, also.

9. The area is 4 × 4 − 6 − 6 − 1/2 = 3 1/2 square units

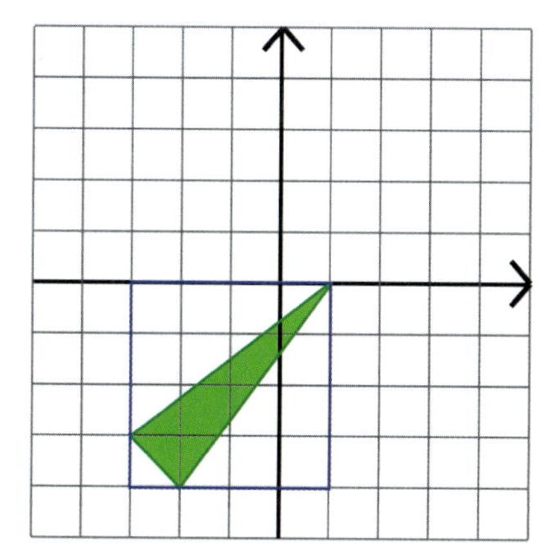

Mixed Review 15, p. 78

1. They put 20 lb of flour into the cellar. Next, 3/8 of the remaining 40 lb is 15 lb. The neighbor got 15 lb.

2. a. 0.072 b. 1.54 c. 25,000 d. 0.0072 e. 0.2 f. 2.1

3. a. 7/10 < 8/10 < 5/6 < 7/8 < 9/10

   b. 11/10 < 9/8 < 7/6 < 12/10 < 10/8

4. a. 0.9 L = 9 dl = 900 cl = 9,000 ml

   b. 2,800 m = 2.8 km = 28,000 dm = 280,000 cm

   c. 56 g = 560 dg = 5,600 cg = 56,000 mg

5.

a. 76 oz = 4.75 lb b. 98 in = 8.17 ft	c. 3.6 gal = 14.4 qt d. 0.483 lb = 7.73 oz	e. 2.67 mi = 14,098 ft f. 5.09 ft = 5 ft 1 in

6.

a. 134 kg $= 134 \text{ kg} \cdot \dfrac{2.2 \text{ lb}}{1 \text{ kg}} = 294.8$ lb
b. 156 in $= 156 \text{ in} \cdot \dfrac{2.54 \text{ cm}}{1 \text{ in}} = 396.24$ cm

7.

a. $0.2m = 6$ **| ÷ 0.2** $m = 6 \div 0.2$ $m = 30$	b. $0.3p = 0.09$ **| ÷ 0.3** $p = 0.09 \div 0.3$ $p = 0.3$	c. $y - 1.077 = 0.08$ **| + 1.077** $y = 0.08 + 1.077$ $y = 1.157$

8. a. See the picture on the right.
   b. The unit rates are:

   2 1/2 squares for <u>1</u> triangle

   2/5 triangles for <u>1</u> square

9.

a. $5 + (-8) = -3$ $(-5) + 8 = 3$	b. $-11 + (-9) = -20$ $9 - 11 = -2$	c. $2 + (-17) = -15$ $-3 - 8 = -11$	d. $2 - (-8) = 10$ $-8 - (-2) = -6$

10. Movements 1. and 2. are shown in the picture. The last movement (the reflection) results in a figure that overlaps the previous one so there is no arrow to show it in the picture.

    The coordinates after the transformations are (−4, 0), (3, 0), (1, −2), and (−3, −2)

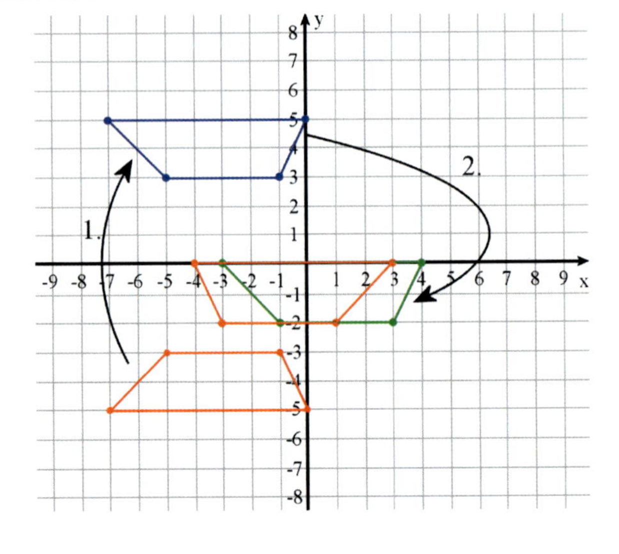

Mixed Review 15, cont.

11. The area is $8 \times 11 \div 2 = 44$ square units.

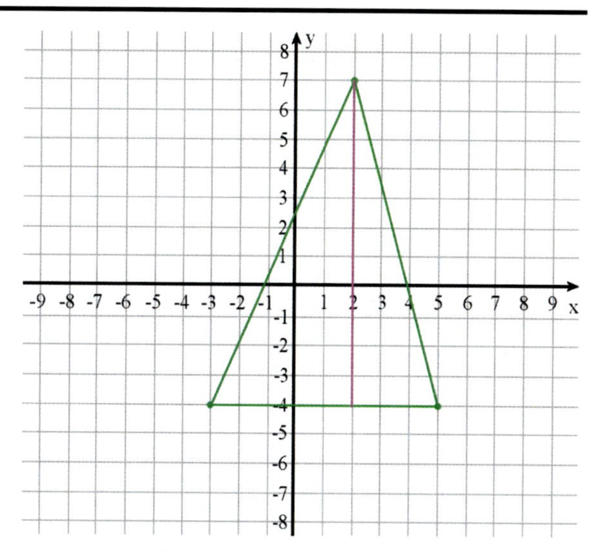

12. a. You can choose any two letters, but a good choice is t for time and d for distance.

b.

time (hours)	0	1	2	3	4	5	6	7	8	9
distance (meters)	0	4	8	12	16	20	24	28	32	36

c. $d = 4t$ or $t = d/4$

d. Time (or t) is the independent variable, because the distance depends on the time and the distance is the dependent variable.

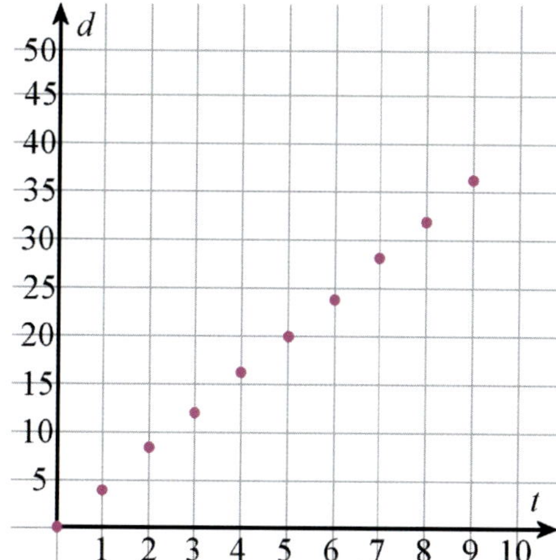

13. a. Dividing a number by 5 is the same as multiplying it by 1/5. Example: $8 \div 5$ is the same as $8 \times 1/5$. Both equal 8/5. Another example: $20 \div 5$ is equal to $20 \times 1/5$. Both are equal to 4.

b. Dividing a number by 2/3 is the same as multiplying it by 3/2. Example: $6 \div (2/3)$ is the same as $6 \times (3/2)$. Both equal 9.

14. a. 5/8 = 63% (exactly 62.5)
b. 6/25 =24%

15.

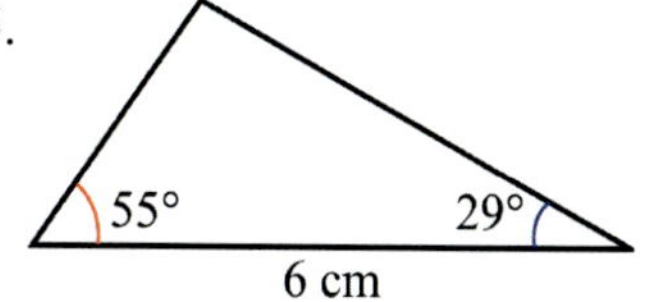

16. First, draw the 66° angle, then measure the two 7.5 cm sides. Or, draw one of the 7.5 cm sides first. Then you can use the compass for the last two sides. The image is not to scale.

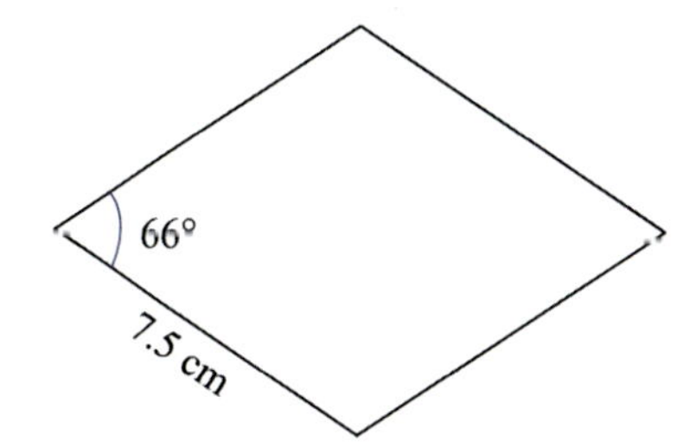

Puzzle corner. a. 1/4 b. 10 c. 6/15 or 2/5

Mixed Review 16, p. 81

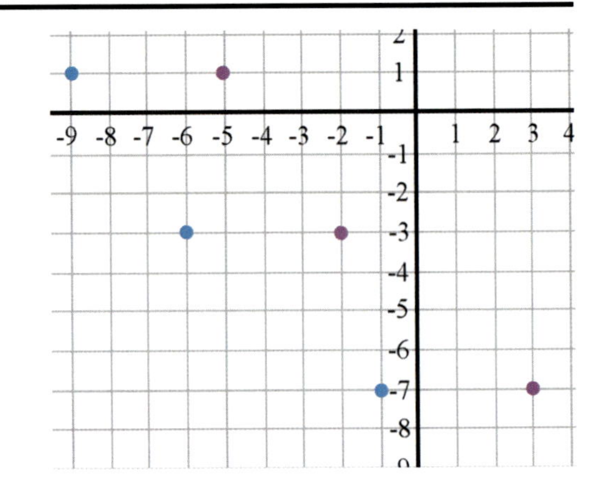

1. $(-5, 1) \rightarrow (-9, 1)$
   $(-2, -3) \rightarrow (-6, -3)$
   $(3, -7) \rightarrow (-1, -7)$

2. $\$180 \div 9 = \$20$; $4 \cdot \$20 = \$80$; $5 \cdot \$20 = \$100$
   Sam got \$80 and Matt got \$100.

3. a. 14 b. 12

4. Answers will vary. Check the student's answers.
   For example, 45, 90, 135, 180, and 225.

5.

a. $4 \times 0.0003 = 0.0012$	b. $0.2 \times 0.3 = 0.06$	c. $0.03 \times 1{,}000 = 30$

6.

a. $0.5x = 30$ $x = 60$	b. $0.01x = 2$ $x = 200$	c. $c + 1.1097 = 3.29$ $c = 2.1803$

7. a. The game pieces are 65 units apart.
   b. Samantha said, "You missed by **60** units!"
   c. She has $2/6 = 33\%$ of her pieces left.

8. The first flash drive costs $0.85 \times \$18 = \$15.30$ after the discount, and the second costs $\$20 \div 5 \times 4 = \$16$. So, the first one is the better deal.

9. Alice has 27 oranges and Michael has 9.
   After giving some to Beatrice, Alice has $90 \div 5 \times 2 = 36$ oranges.
   She gave 1/4 of those, or 9 oranges to Michael.

10.

$$\frac{\$250}{100 \text{ sq. ft.}} = \frac{\$500}{200 \text{ sq. ft.}} = \frac{\$1{,}250}{500 \text{ sq. ft.}} = \frac{\$5{,}000}{2{,}000 \text{ sq. ft.}} = \frac{\$6{,}000}{2{,}400 \text{ sq. ft.}}$$

11.

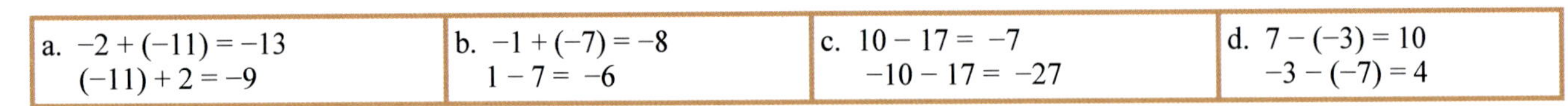

a. $-2 + (-11) = -13$ $(-11) + 2 = -9$	b. $-1 + (-7) = -8$ $1 - 7 = -6$	c. $10 - 17 = -7$ $-10 - 17 = -27$	d. $7 - (-3) = 10$ $-3 - (-7) = 4$

12.

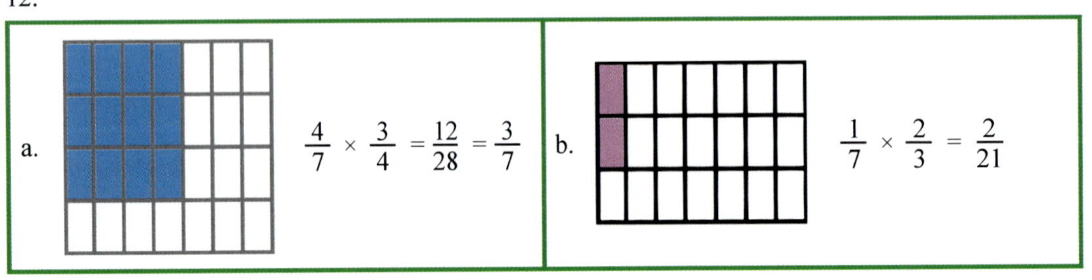

a. $\frac{4}{7} \times \frac{3}{4} = \frac{12}{28} = \frac{3}{7}$

b. $\frac{1}{7} \times \frac{2}{3} = \frac{2}{21}$

Statistics Review, p. 83

1. a. Yes.
   b. No. Change it for example to: How many pages are in the 7th grade math books? Or, How many pages are in the problem-solving books?

2. a. About <u>97%</u> of the population of Norway use the Internet.
   b. About <u>84%</u> of the population of United Kingdom use the Internet.
   c. So, there are about <u>15.0</u> million internet users in the Netherlands.
   (Multiply $0.89 \times 16{,}847{,}000$ and round to the nearest tenth of a million.)
   d. So, there are about <u>4.6</u> million internet users in Finland.
   (Multiply $0.88 \times 5{,}260{,}000$ and round to the nearest tenth of a million.)

3. Mean: 7.6. Median: 7.5. Mode: 7. (The original data is 4, 6, 6, 7, 7, 7, 7, 7, 7, 8, 8, 8, 8, 9, 9, 9, 10, 10.)
   b. It is more or less bell-shaped, though slightly left-tailed (left-skewed).

4. a. minimum 2
   1st quartile 5.5
   median 7
   3rd quartile 8
   maximum 12
   interquartile range 2.5

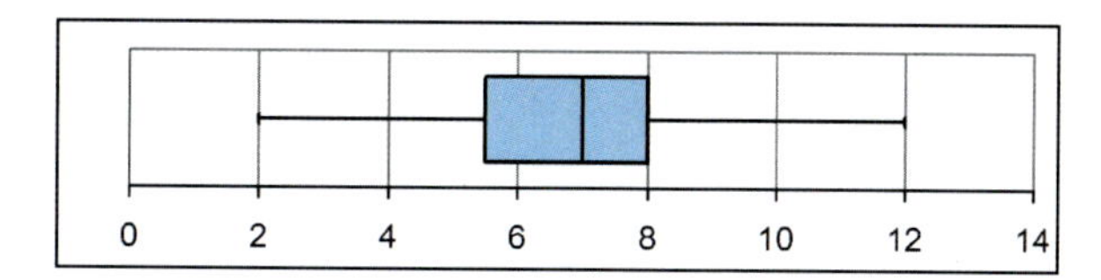

   b. Answers vary. It could be the ages of some group of children, such as children playing at the playground.

5. a.

Stem	Leaf
6	4 6
7	5 7 8
8	2 4 4 5
9	0
10	
11	2

   b. 82
   c. 48
   d. 112
   e. Without the outlier, it is J-shaped.

6. a. about 1/2 of the people
   b. It is J-shaped.
   c. The median is, because it falls within the highest bar, whereas the mean does not.

7. a.

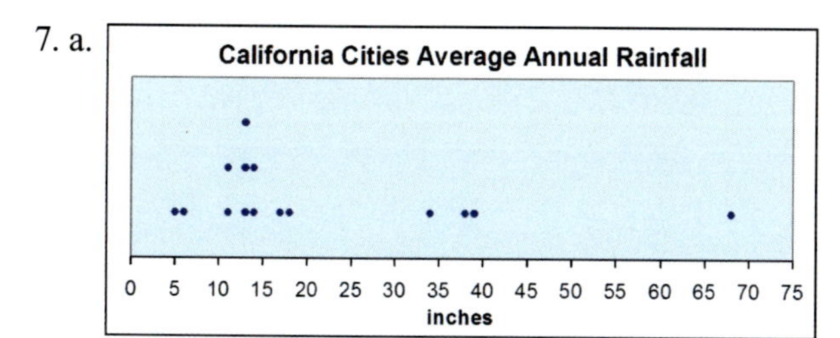

   b. The data is spread out a lot.
   c. The distribution is more or less bell-shaped and it has an outlier.
   d. Median: 14 inches. Note that mean is not a good measure of center to use here because of the outlier. The outlier would throw the mean off.

   e.

Rainfall (in)	Frequency
5 - 20	11
21 - 36	1
37 - 52	2
53 - 69	1

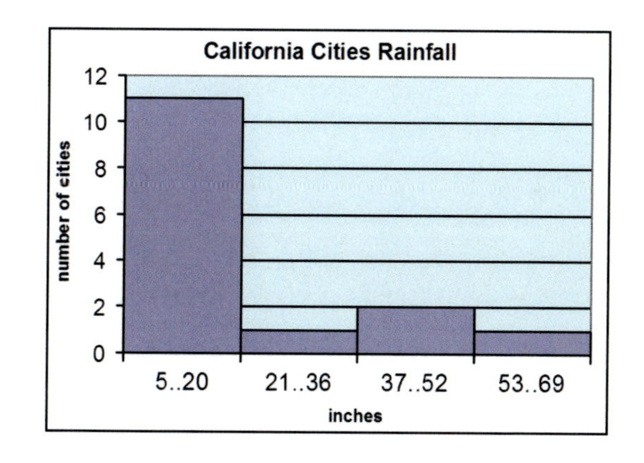

Statistics Test, p. 86

1. a. mean 15.3 median 15 mode 15 range 9
   b. mean (none) median (none) mode horse, dog range (none)

2. a., b. and c.

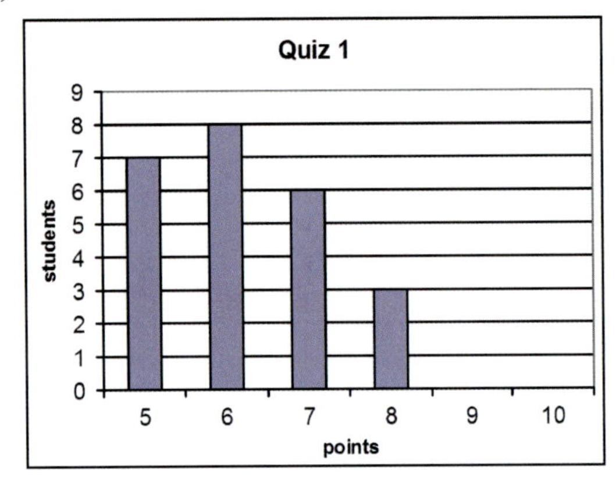

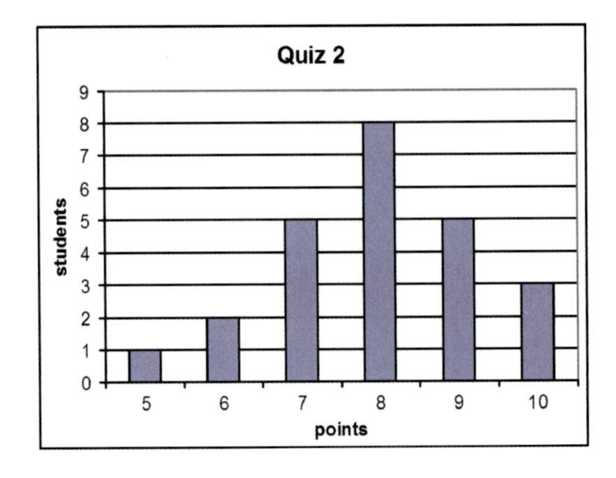

Quiz 1:

right-tailed or right-skewed

mean 6.21
median 6
mode 6

Quiz 2:

fairly bell-shaped

mean 7.96
median 8
mode 8

c. Any of the three measures of center can be used. All three are listed above.
d. Quiz 2

3. a.

Stem	Leaf
11	4 9
12	0 1 2 5 7 7 8
13	0 2

b. The median is 125.
c. The interquartile range is 128 − 120 = 8.

4.

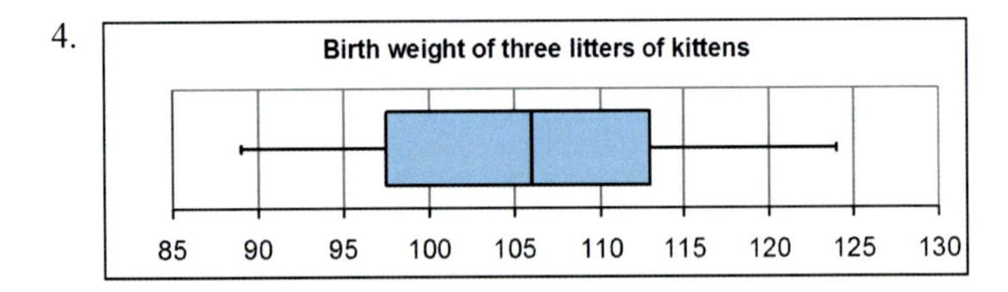

Five-number summary:
Minimum 89
1st quartile 97.5
Median 106
3rd quartile 113
Maximum 124

Mixed Review 17, p. 88

1. a. 3 b. 16

2. a. 12 b. 24

3.

a. GCF of 72 and 12 is 12	b. GCF of 42 and 66 is 6
12 + 72 = 12 (1 + 6)	42 + 66 = 6 (7 + 11)

4. a.

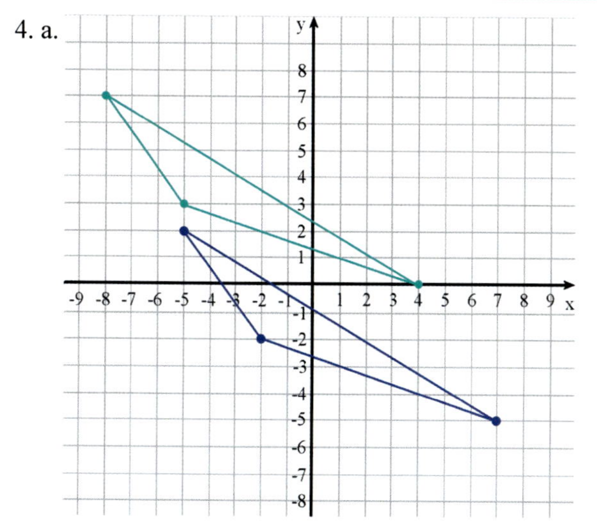

b. (−5, 2), (−2, −2) , and (7, −5)

5. a. $19s = 304$. $s = 304 \div 19 = 16$.
The other side is 16 m.

b. $5w = 6.7$. $w = 6.7 \div 5 = 1.34$.
Each book weighed 1.34 kg.

6. a. $7z^3$
b. $240ab$
c. $4x + 2$
d. $3t + 6$

7. a. (60 + 66 + 110)/132 = 1 104/132 = 1 26/33

b. 3 55/60 − 30/60 + 15/60 = 3 40/60 = 3 2/3

8. (2/3) × (9/10) = 18/30 = 3/5 of the whole pizza.

9. The piglet gains 12 × 7 1/3 oz = 88 oz = 5 lb 8 oz.
Its final weight is 3 lb 4 oz + 5 lb 8 oz = 8 lb 12 oz.

10. 5 3/4 in ÷ 4 = 23/4 in ÷ 4 = 23/16 in = 1 7/16 in.

11. a. 9
b. 3
c. 0
d. 28
e. 7

12. a. $^{-}12 + 7 = {}^{-}5$
b. $2 - 8 = {}^{-}6$

13. Check the student's pictures. The product of the two sides that are perpendicular to each other should be 16. For example, the two perpendicular sides could measure 4 in. and 4 in, or 2 in. and 8 in., or 5 1/3 in. and 3 in., etc.

14. Sail: 3 × 6 ÷ 2 = 9 square units.
Boat: 1 + 16 + 1 = 18 square units.
Total: 27 square units.

15. 19 in × 11 in − 10 in × 5.5 in ÷ 2
= 209 in^2 − 27.5 in^2 = 181.5 in^2.

16. a. The three warmest months are June, July, and August. The three coldest months are January, February, and December.
b. 27°C − 2°C = 25°C
c. In August, 9 degrees. In January, also 9 degrees.

Puzzle corner.
a. Jerry "flipped" the first fraction, or uses the reciprocal of the first fraction, and then multiplied.
b. Emily ignored the whole-number part of the mixed numbers.

Mixed Review 18, p. 91

1. Perimeter = 13.7 cm + 22.38 cm + 17.2 cm = 53.28 cm
   Area = 22.38 cm × 0.97 cm ÷ 2 = 10.8543 cm^2

2. a. It is 3,696 feet from Ben's home to his workplace (0.7 × 5,280 ft = 3,696 ft).
   b. Ben walks to work 4 days a week.
   Ben walks 4 × 48 × 3,696ft = 709,632 ft, which is 709,632 ft ÷ 5,280 ft = 134.4 miles.

3.

	m	dm	cm	mm
a. 7.82 m	7.82	0.782	0.0782	0.00782
b. 109 mm	0.109	1.09	10.9	109 mm

4. a. \$17.54 ÷ 3 = \$5.847
   b. 2.4 ÷ 0.05 = 48

5. a. The volume of the shoe box is 5,400 cm^3

   b.

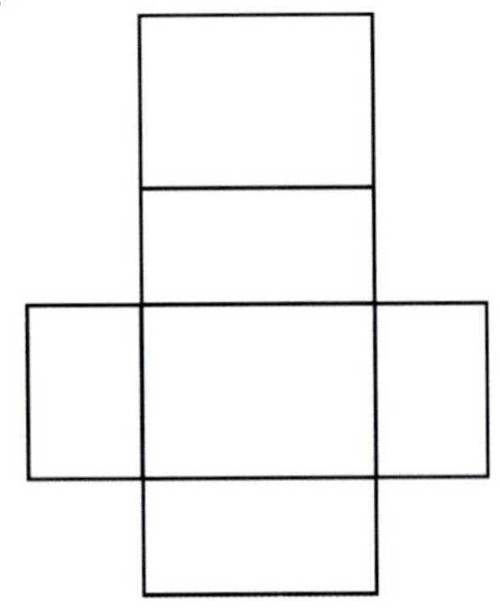

   c. Surface area = 2(18 × 25) + 2(12 × 25) + 2(12 × 18) = 1,932 cm^2

6. a. $x^5 - 5$
   b. $(2 - x)^3$
   c. $2(10 + y)$
   d. $\frac{s-2}{s^2}$

7. 32 − 5 = 27

8. a. $56x + 14 = 14(4x + 1)$
   b. $18u + 60 = 6(3u + 10)$

9.

a.	b.	c.
$y \div 50 = 60 \cdot 2$	$3x - x = 3 + 7$	$7x = 50$
$y \div 50 = 120$	$2x = 10$	$x = 50 \div 7$
$y = 120 \cdot 50$	$x = 10 \div 2$	$x = 7\ 1/7$
$y = 6{,}000$	$x = 5$	

10. {19, 21, 23}

11.

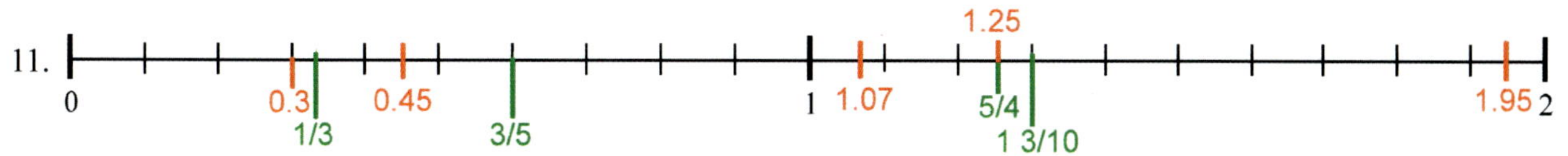

12.

a. $\frac{5}{4} = 1.25$	b. $\frac{6}{7} = 0.857$	c. $\frac{19}{16} = 1.188$

13. 548.386 cm^2

In centimeters, the puzzle measures $8.5 \cdot 2.54 = 21.59$ cm by $10 \cdot 2.54 = 25.4$ cm. Its area is then $21.59 \cdot 25.4 = 548.386$ cm^2.

14. The oats cost \$2.53. This is easiest to calculate by changing 2 3/4 lb into the decimal 2.75 lb: \$0.92 · 2.75 = \$2.53.

15.

Stem	Leaf
14	0 5 8
15	0 2 5 8 9
16	0 2 2 3 3
17	2 5
18	
19	0

End-of-the-Year Test, p. 95

Please see the test for grading instructions.

The Basic Operations

1. a. $2{,}000 \div 38 = 52$ R4. There will be 52 bags of cinnamon.

2. a. $2^5 = 32$ b. $5^3 = 125$ c. $10^7 = 10{,}000{,}000$

3. a. 70,200,009 b. 304,500,100

4. a. 6,300,000 b. 6,609,900

Expressions and Equations

5. a. $s - 2$ b. $(7 + x)^2$

c. $5(y - 2)$ d. $\dfrac{4}{x^2}$

6. a. $40 - 16 = 24$

b. $\dfrac{65}{5} = 13 \cdot 3 = 39$

7. a. $\$50 - 2m$ or $\$50 - m \cdot 2$
b. s^2

8. $z + z + 8 + x + x + x = 2z + 3x + 8$ or $3x + 2z + 8$ or $2z + 8 + 3x$

9. $6(s + 6)$ or $(s + 6 + s + 6 + s + 6 + s + 6 + s + 6 + s + 6$. It simplifies to $6s + 36$.

10. $6b \cdot 3b = 18b^2$

11. a. $3x$ b. $14w^3$

12. a. $7(x + 5) = 7x + 35$ b. $2(6p + 5) = 12p + 10$

13. a. $\underline{2}(6x + 5) = 12x + 10$ b. $5(2h + \underline{6}) = 10h + 30$

14.

a.	b.
$\dfrac{x}{31} = 6$	$a - 8.1 = 2.8$
$x = 6 \cdot 31$	$a = 2.8 + 8.1$
$x = 186$	$a = 10.9$

15. $y = 2$

16. $0.25 \cdot x = 16.75$ OR $25x = 1675$. The solution is $x = 67$ quarters.

17. a. $p \le 5$
The variable students use for "pieces of bread" may vary.
b. $a \ge 21$
The variable students use for "age" may vary.

18.

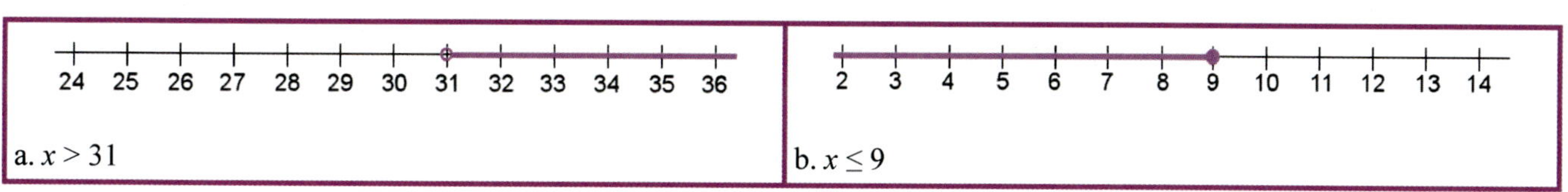

a. $x > 31$ b. $x \le 9$

19. a.

t (hours)	0	1	2	3	4	5	6
d (km)	0	80	160	240	320	400	480

b. See the grid on the right.
c. $d = 80t$
d. t is the independent variable

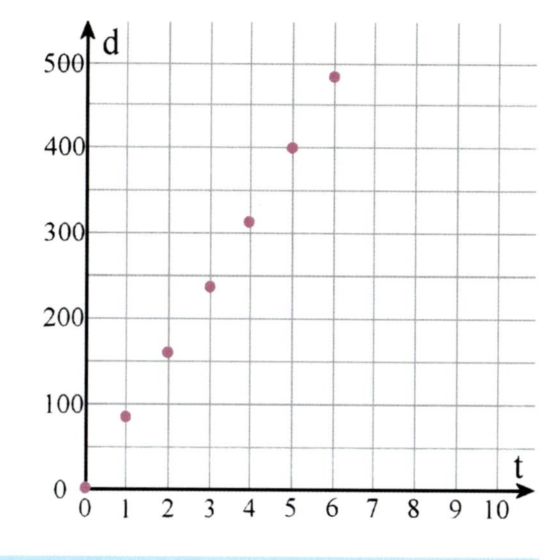

Decimals

20. a. 0.000013 b. 2.0928

21. a. $\frac{78}{100,000}$ b. $2 \frac{302}{1,000,000}$

22. 0.0702

23. a. 8 b. 0.00048

24. a. Estimate: $7 \times 0.006 = 0.042$
b. Exact: $7.1 \times 0.0058 = 0.04118$

25. $1.5 + 0.0022 = 1.5022$

26. a. 90,500 b. 0.0024

27. a. $175 \div 0.3 = 583.333$

b. $\frac{2}{9} = 0.222$

28. a. Estimate: $13 \div 4 \times 3 = (3\ 1/4) \times 3 = \$9.75$
b. Exact: \$9.69

29. $(3 \times \$3.85 + \$4.56) \div 2 = \$8.06$

Measuring Units

30. a. 178 fl. oz. = 5.56 qt b. 0.412 mi. = 2,175.36 ft c. 1.267 lb = 20.27 oz

31. 0.947 mile

32. You can get 10 six ounce serving and have 4 ounces left over.

33. It is about \$6.65 per pound. To calculate the price per pound, simply divide the cost by the weight in pounds. A pack of 36 candy bars weighs 36 × 1.55 oz = 55.8 oz = 3.4875 lb. Now simply divide the cost of those candy bars by their weight in pounds to get the price per pound: \$23.20 ÷ 3.4875 lb = \$6.652329749103943 / lb.

34. a. 39 dl = 3.9 L

			3	9		
kl	hl	dal	l	dl	cl	ml

b. 15,400 mm = 15.4 m

		1	5	4	0	0
km	hm	dam	m	dm	cm	mm

c. 7.5 hm = 75,000 cm

	7	5	0	0	0	
km	hm	dam	m	dm	cm	mm

d. 597 hl = 59,700 L

5	9	7	0	0			
	kl	hl	dal	l	dl	cl	ml

e. 7.5 hg = 0.75 kg

0	7	5				
kg	hg	dag	g	dg	cg	mg

f. 32 g = 3,200 cg

		3	2	0	0	
kg	hg	dag	g	dg	cg	mg

35. a. Twenty-four bricks will cover the span of the wall. 5150 mm ÷ 215 mm = 23.953488.
b. Twenty-four bricks will still cover the span of the wall. 5150 mm ÷ 216 mm = 23.842593.

Ratio

36. a.

b. 10:15 = 2:3

37. a. 3,000 g:800 g = 15:4 b. 240 cm:100 cm = 12:5

38. a. \$7:2 kg b. 1 teacher per 18 students

39. a. \$4 per t-shirt. b. 90 miles in an hour

40. a. You could mow 20 lawns in 35 hours.
b. The unit rate is 105 minutes per lawn (or 1 h 45 min per lawn).

Lawns	4	8	12	16	20
Hours	7	14	21	28	35

41. Mick got \$102.84. \$180 ÷ 7 × 4 = \$102.84.

42.

a. $7.08 \text{ mi} = 7.08 \text{ mi} \cdot \dfrac{1.6093 \text{ km}}{1 \text{ mi}} \approx 11.394 \text{ km}$

b. $4 \text{ L} = 4 \text{ L} \cdot \dfrac{1 \text{ qt}}{0.946 \text{ L}} \approx 4.23 \text{ qt}$

Percentage

43.

a. 35% = $\frac{35}{100}$ = 0.35	b. 9% = $\frac{9}{100}$ = 0.09	c. 105% = $1\frac{5}{100}$ = 1.05

44.

	510
1% of the number	5.1
5% of the number	25.5
10% of the number	51
30% of the number	153

45. The discounted price is \$39. You can multiply 0.6 × \$65 = \$39, or you can find out 10% of the price, which is \$6.50, multiply that by 4 to get the discount (\$26), and subtract the discounted amount.

46. The store had 450 notebooks at first. Since 90 is 1/5 of the notebooks, the total is 90 ×5 = 450.

47. She has read 85% of the books she borrowed from the library. 17/20 = 85/100 = 85%.

Prime Factorization, GCF, and LCM

48. a. 3 × 3 × 5 b. 2 × 3 × 13 c. 97 is a prime number

49. a. 8 b. 18

50. a. 2 b. 15

51. Any three of the following numbers will work: 112, 140, 168, 196

52.

a. GCF of 18 and 21 is 3. 18 + 21 = 3· 6 + 3 ·7 = 3(6 + 7)
b. GCF of 56 and 35 is 7. 56 + 35 = 7(8 + 5)

Fractions

53. a. 4 b. 2 1/12 c. 5 3/5

54. $3\frac{2}{3} \div \frac{3}{5} = 6\frac{1}{9}$

55. Answers will vary. Please check the student's work.
Example: There was 1 3/4 pizza left over and three people shared it equally. Each person got 7/12 of a pizza.

56. There are ten servings. (7 1/2) ÷ (3/4) = (15/2) ÷ (3/4) = (15/2) × (4/3) = 60/6 = 10.

57. 63 8/9 square feet.
The area of the room is (12 1/2) × (15 1/3) = (25/2) × (46/3) = 25 × 23/3 = 575/3 = 191 2/3 square feet.
One-third of that is (191 2/3) × (1/3) = 574/9 = 63 8/9.
Or, you can first divide one of the dimensions by three, and then multiply to find the area.

58. 4 13/20 inches and 3 1/10 inches or 4.65 inches and 3.1 inches.

The ratio of 3:2 means the two sides are as if three "parts" and two "parts", and the total perimeter is 10 of those parts. Therefore, one part is 15 1/2 in. ÷ 10 = 15.5 in. ÷ 10 = 1.55 inches. The one side is three times that, and the other is two times that. So, the sides are 4.65 in. and 3.1 in. If you use fractions, you get (15 1/2 in.) ÷ 10 = (31/2 in.) ÷ 10 = 31/20 in., and the two sides are then 3 × 31/20 in. = 93/20 in. = 4 13/20 in. and 2 × 31/20 in. = 62/20 in. = 3 1/10 in.

Integers

59. a. > b. >

60. a. −7°C > −12°C. b. $5 > −$5.

61. a. The difference is 23 degrees. b. The difference is 12 degrees.

62. a. −7 b. |−6| = 6 c. |5| = 5 d. |−6| = 6

63. a.- c See the grid on the right.
d. 6 × 10 ÷ 2 = 30
The area of the resulting triangle is 30 square units.

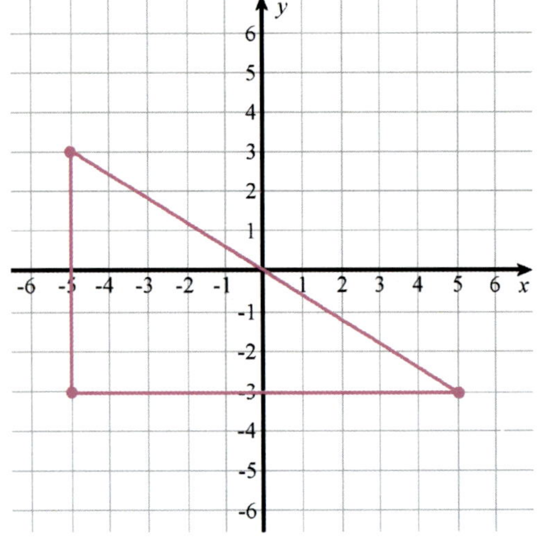

64. a. −2 + 5 = 3

b. −2 − 4 = −6

c. −1 − 5 = −6

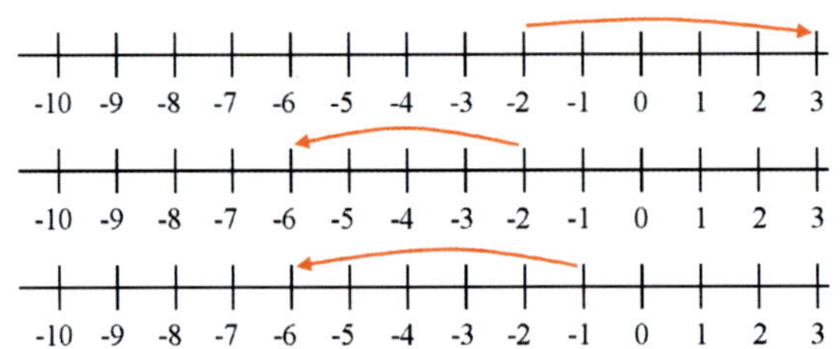

65. a. That would make his money situation to be −$4.

$10 −$14 = −$4
OR
$10 + (−$14) = −$4

b. Now he is at the depth of −3 m.

−2 m −1 m = −3 m
OR
−2 m + (−1 m) = −3 m

Geometry

66. The area is 4 × 3 ÷ 2 = 6 square units.

67. Answers may vary. The base and altitude of the parallelogram could be for example 5 and 3, or 3 and 5, or 6 and 2 1/2.

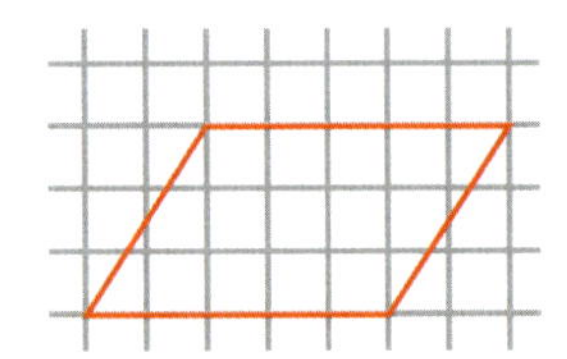

68. Divide the shape into triangles and rectangles, for example like this:

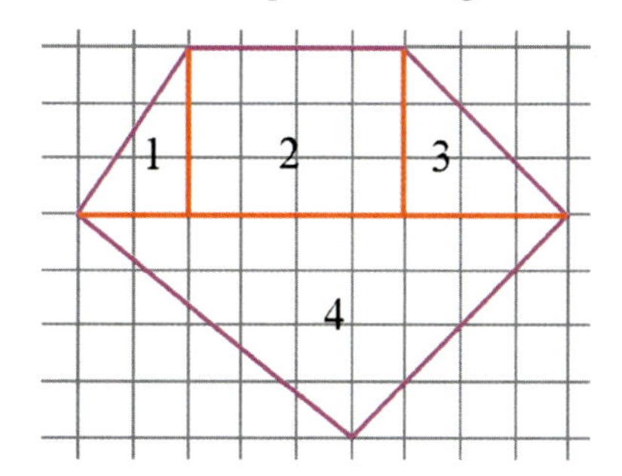

The areas of the parts are:

triangle 1: 3 square units
rectangle 2: 12 square units
triangle 3: 4.5 square units
triangle 4: 18 square units

The overall shape (pentagon): 37.5 square units

69. It is a trapezoid. To calculate its area, divide it into triangles and rectangle(s).

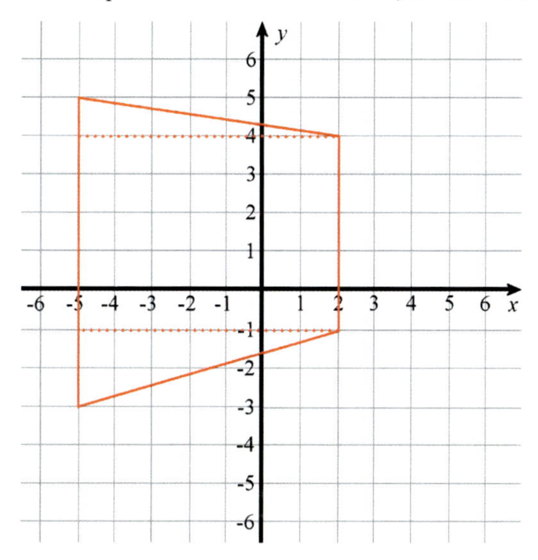

The area is: 3.5 + 35 + 7 = 45.5 square units

70. It is a triangular prism. Some possible nets are shown below:

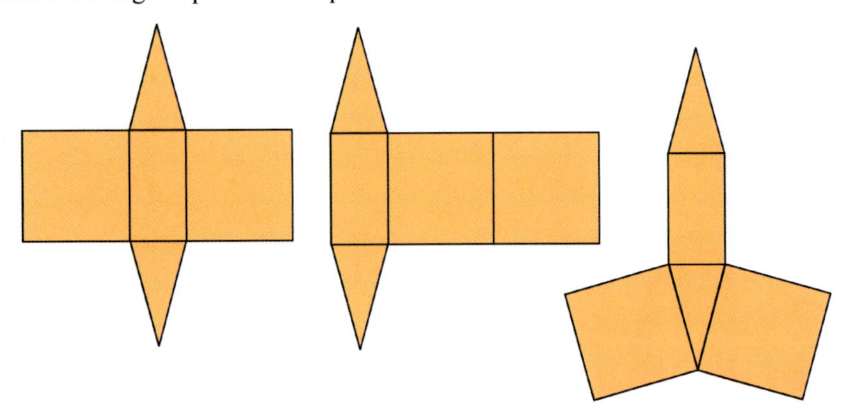

71. a. It is a rectangular pyramid.

b. The rectangle has the area of 300 cm^2. The top and bottom triangles: 2×20 cm $\times$ 11.2 cm $\div 2 = 224$ cm^2. The left and right triangles: 2×15 cm $\times$ 13 cm $\div 2 = 195$ cm^2. The total surface area is 719 cm^2.

72. The volume of each little cube is (1/2 cm) $\times$ (1/2 cm) $\times$ (1/2 cm) = 1/8 cm^3.

a. 18 $\times$ (1/8) cm^3 = 18/8 cm^3 = 9/4 cm^3 = 2 1/4 cm^3.

b. 36 $\times$ (1/8) cm^3 = 36/8 cm^3 = 9/2 cm^3 = 4 1/2 cm^3.

73. a. 1 3/4 in $\times$ 8 1/2 in $\times$ 6 in = (7/4) in $\times$ (17/2) in $\times$ 6 in = (119/4) $\times$ 6 in^3 = (29 3/4) $\times$ 3 in^3 = 87 9/4 in^3 = 89 1/4 in^3. This calculation can also be done (probably quicker) by using decimals: 1.75 in $\times$ 8.5 in $\times$ 6 in = 89.25 in^3.

b. Imagine you place the boxes in rows, standing up, so that the height is 6 inches. Then we can stack two rows on top of each other, since the height of the box is 1 ft or 12 inches. The width of each box is 1 3/4 in, and 6 boxes fit in the space of 1 ft., because 6 $\times$ (1 3/4 in) = 6 18/4 in = 10 1/2 in. Since the last dimension is over 8 inches, we cannot fit but one row. So, we can fit two rows of 6 boxes, stacked on top of each other, or a total of 12 boxes.

Statistics

74. a. See the plot on the right.

b. The median is 68.5 years.

c. The first quartile is 63, and the third quartile is 75.5. The interquartile range is thus 12.5 years.

Stem	Leaf
5	5 9
6	1 2 4 5 5 8 9
7	0 2 4 7
8	3 9
9	4

75. a. It is right-tailed or right-skewed. You can also describe it as asymmetrical.

b. Median. Mean is definitely not the best, because the distribution is so skewed. Without seeing the data itself, we cannot know if mode would work or not - it may not even exist, since typically for histograms, the data is very varied numerically and has to first be grouped.

76. a.

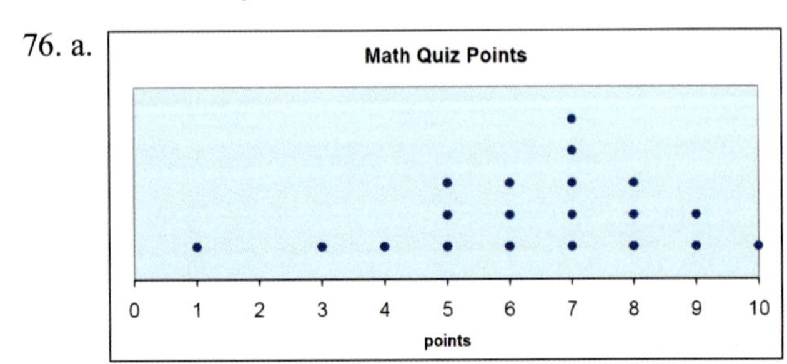

b. It is fairly bell-shaped but is somewhat left-tailed or left-skewed. You can also say it is asymmetrical.

c. The data is spread out a lot.

d. Any of the three measures of center works. Mean: 6.4. Median: 7. Mode: 7.

Made in the USA
Middletown, DE
19 August 2019